QUANTUM MODELLING OF ECONOMIC AND FINANCIAL SYSTEMS

This concise and self-contained book opens completely novel areas of research by directly implementing concepts from quantum physics into areas of social science. It constructs compelling arguments originating from fundamental concepts in physics and the philosophy of science, including key developments in economics and finance, then surveys the important work which has been performed to date through applying the formalism of quantum mechanics to decision-making and finance. The book is accessible to graduate students and researchers in social science and physics, as well as avid interdisciplinary readers.

This title is part of the Flip it Open Programme and may also be available Open Access. Check our website Cambridge Core for details.

VIKRAM ATHALYE is Associate Professor of Physics at Cummins College of Engineering for Women Pune, India and a TEDx speaker. His research currently focuses on bridging physics and the social sciences. Previously he has investigated the foundations of physics by means of quantum information methods. Vikram has published in several leading journals of physics, including *Physical Review Letters*.

EMMANUEL HAVEN holds the Dr. Alex Faseruk Chair in Financial Management at Memorial University in Canada. He has published in diverse areas ranging from the social sciences such as psychology, economics, and operations research, to biology, physics, and applied mathematics. He has published several books including *Quantum Social Science* (Cambridge University Press, 2013). He is the co-founding editor of the new journal *Quantum Economics and Finance* (Sage).

QUANTUM MODELLING OF ECONOMIC AND FINANCIAL SYSTEMS: SELECTED TOPICS

VIKRAM ATHALYE
Cummins College of Engineering for Women, Pune, India

EMMANUEL HAVEN
Memorial University of Newfoundland and Labrador

CAMBRIDGE
UNIVERSITY PRESS

CAMBRIDGE
UNIVERSITY PRESS

Shaftesbury Road, Cambridge CB2 8EA, United Kingdom

One Liberty Plaza, 20th Floor, New York, NY 10006, USA

477 Williamstown Road, Port Melbourne, VIC 3207, Australia

314–321, 3rd Floor, Plot 3, Splendor Forum, Jasola District Centre, New Delhi – 110025, India

103 Penang Road, #05–06/07, Visioncrest Commercial, Singapore 238467

Cambridge University Press is part of Cambridge University Press & Assessment, a department of the University of Cambridge.

We share the University's mission to contribute to society through the pursuit of education, learning and research at the highest international levels of excellence.

www.cambridge.org
Information on this title: www.cambridge.org/9781108844666

DOI: 10.1017/9781108953177

When citing this work, please include a reference to the DOI 10.1017/9781108953177

First published 2026

A catalogue record for this publication is available from the British Library

A Cataloging-in-Publication data record for this book is available from the Library of Congress

ISBN 978-1-108-84466-6 Hardback

To my parents, Usha & Sharad and
To my wife and son, Vrushali & Arin
— V. A.

To 'Popke', Natty, Braunty and kids: Hazel-Noisette,
Holly-Eloise and Sammy, Jane
— E. H.

Contents

Foreword 1 William Sulis

As many of us move closer to retirement, financial stability becomes ever more prominent in our minds. Reminiscence is often through rose-coloured glasses, but it seems as though the years leading up to and following the financial collapse of 2008 have been unparalleled in the degree of instability observed in the market. There has been bubble after bubble, large fluctuations, and a pervasive sense of insecurity. There have been global shocks to the market, such as the Covid-19 pandemic of 2020–2023, but even local events such as the Russian invasion of Ukraine in 2022 are having global consequences. The use of technology, meant to facilitate trading on stock exchanges, has seemingly amplified the effect of these shocks, making the situation worse, not better. Resilience has given way to fragility.

Several years ago, I attended a conference on complex systems, and one of the speakers there represented the financial arm of the US government. They came to the meeting seeking out collaborations with complex systems theorists in the hope of developing economic models that were more closely linked to the reality of our markets. They pointed out that not only did the then current economic models fail to predict the 2008 collapse, but the models also actually claimed that such a collapse was impossible. They said that the models were not merely wrong, but that their fundamental assumptions were wrong, and that a new approach was needed.

I think that one of the first major advances in economic theory came with the development of behavioural economics. For the first time, decision-making was not treated as a philosophical subject grounded in ideal theories of logic, optimality, rationality, and epistemology (LORE), but rather as an empirical subject, grounded in observations of actual human agents making decisions. Initially, emphasis was placed on the biases and irrationality of human decision-making, as if our departure from the ideals of LORE was a serious failing on our part. Over time, I think a more nuanced viewpoint has emerged, particularly within the literature on collective intelligence, which has demonstrated the adaptive utility of the ability to adopt different decision-making strategies in different contexts involving differing constraints. Theories based on LORE are theories based on ideas of perfection – perfect knowledge, perfect reasoning, perfect evaluations of outcomes, perfect conditions. None of these hold in the real world in which people and animals are forced to make decisions with limited, imperfect knowledge, limited experience with formal logic (and using contextual rather than propositional facts), limited ability to evaluate outcomes and to predict future events, all while under significant time and resource constraints. The very survival of people and animals relies on the effectiveness of their decisions, and despite all of these limitations, both manage fairly well most of the time. LORE fails under these conditions. Far from being irrational, these nonrational decision-making strategies are exploited 'on the fly' and are 'good enough' for dealing with these situations.

The assumptions of LORE make for beautiful mathematics, but the connection of these mathematical results to the real world is tenuous at best. Markets do not tend toward equilibria – fluctuations are not Gaussian but rather follow power-law scaling typical of the fractal structure of a complex system. Fluctuations are not merely random noise but are an intrinsic feature of market dynamics. As we have already experienced, the consequences of using models which are ill fitted to the dynamics of the market can be devastating and result in real-world suffering.

The next major advance in economic theory was the introduction of ideas from nonlinear dynamics and complex systems theory. Spearheaded by the Santa Fe Institute and further advanced by groups such as the New England Complex Systems Institute and the Society for Chaos Theory in Psychology and the Life Sciences, there has been a major effort to develop economic models and analyses of experiments which are more in tune with the complexity of real systems, and with empirical observations, rather than idealized models based in LORE. Sadly, many of these ideas, while expressed in popular media, have not had much impact on mainstream economic theory. This book offers a third advance in economic thinking by proposing the modelling of the behavioural aspects of the agents of economy in a physics-analogous manner.

Logic, Optimality, Rationality, and Epistemology aside, economic theory has, in its probabilistic aspects, been grounded upon Kolmogorov's theory of probability. Empirical studies of decision-making have shown for decades that people often make probabilistic evaluations which do not conform to Kolmogorov's rules. This has usually been explained as an error in human reasoning. A few errors are plausible, but these deviations appear to be robust across a wide range of subjects. Such systematic departures demand explanation. It is easy to hold to Kolmogorov theory (which is compatible with LORE) and blame these deviations on human foibles. The discovery of quantum mechanics, however, brought with it an entirely new and unexpected probability theory in the form of Born's rule for determining probabilities from wave functions. This introduced interference terms into the usual sum rule for decomposing a probability into a sum of conditional probabilities.

This was the first example of a non-Kolmogorov probability and since then two additional probability models have been proposed: Khrennikov's Contextual Probability Theory (Khrennikov) and Dzhafarov's Contextuality by Default Theory (Dzhafarov). Both have been applied in psychological and economic settings. Both approaches take into account another feature of real-world systems: contextuality. At its most basic, contextuality refers to the observation that the form of random variables depends upon the context within which they are measured. The same measurement tool, applied to the same subjects, but in different contexts, may result in different random variables (i.e. probabilities) being obtained. This means that it is not always possible to arbitrarily combine results obtained in different contexts and to form joint probability distributions from which calculations of quantities of interest may be obtained. The study of this form of contextuality is vital to the advancement of the social, biomedical, and economic sciences.

There is a deeper form of contextuality, however, which relates to the kinds of correlations which are possible, and which are reflected in the violation of certain inequalities (so-called Bell inequalities) formed by combining various correlations. Though it is often claimed that only quantum systems violate these inequalities, there is evidence for their violation in social science settings as well.

This book offers a detailed examination of the application of both mathematical formalism (non-Kolmogorov probability) as well as some of the conceptual underpinnings of quantum mechanics, to the study of economics, in particular to finance. I think that the main contributions of quantum mechanics to the social sciences lie in demonstrating the fundamental stochasticity of the natural world, the existence of non-Kolmogorov probability, the importance of contextuality, and the discrete nature of exchanges between entities.

This book not only presents a detailed discussion on how analogical arguments and analogical reasoning could be used to develop a quantum physics–analogous approach to the construction of economics and finance models, but also demonstrates in detail how such an approach would work in a real financial context. One of the most interesting aspects of this book is a clear and transparent discussion concerning the application of the ideas of causality and ensemble from physics to the social sciences. The other one is a balance between the epistemological and the ontological aspects of the interpretation of quantum theory when applied to the social sciences.

I am sure that the tools of physics – symmetry and its transformations, scaling, nonlinear dynamics, interacting particle systems, transience, and emergence, to name a few – when employed in the social sciences in an analogous manner, will be of great value.

I hope that this book will inspire readers themselves to venture forward into this new and exciting world of quantum-inspired economics and help to build economic theories appropriate for the twenty-first century.

William Sulis, BSc, MD, MA, FRCPC, PhD, PhD

Foreword 2 Andrei Khrennikov

The past 20–30 years were characterized by the tremendous development of quantum information theory. Sometimes the information restructuring of quantum physics is called the second quantum revolution. This revolutionary and accelerating propagation of ideas of information theory to quantum physics has not only impacted the quantum technologies (such as quantum computing, cryptography, and teleportation), but has also stimulated applications of the mathematical formalism and methodology of quantum theory in the social sciences. Such applications to cognition, psychology, decision-making, social and political sciences, economics, and finance are known as quantum-like models. This is a flowering area of modern multidisciplinary science unifying the efforts of researchers both from natural science and humanities. This is a good place to remark that Google search generates 213 000 000 results on 'quantum-like modelling.'

Nowadays quantum-like models are widely applied to describe mathematically the brain's functioning, cognition, and psychological effects, such as conjunction, disjunction, question order, and response replicability effects. Recently they have started to diffuse into social and political science, economics, and finance. In quantum-like research, the emphasis was on exploring the quantum mathematical apparatus – the operator calculus in complex Hilbert space. Quantum states are represented by density operators describing the statistical features of ensembles of similar systems. Quantum observables are represented by Hermitian operators. The probability distributions of outcomes of observations are given by the Born rule.

The main distinguishing feature of this formalism is the noncommutativity of operators representing some observables, so-called incompatible observables. Such observables cannot be represented as random variables on the same Kolmogorov probability space, that is, classical probability theory is nonapplicable. The existence of such observables in physics is a direct consequence of the quantization procedure describing transition from the classical mechanics on the phase space to quantum mechanics. In applications of quantum theory to the social sciences, one cannot proceed in this way, since, say for cognition as well as economics and finance, the classical Hamiltonian formalism does not exist. Therefore, existence of incompatible observables is a complex methodological problem which should be solved not via the mathematical calculations, but via a deep analysis of the origin of observables and their meaning in the corresponding area of science. In psychology and decision-making, some progress in this direction was made, in particular, via analysis of disjunction effect that is mathematically described as interference of probabilities. In economics and finance the situation is even more complex. This stimulates the interest in the analysis of the conceptual background of quantum-like modelling.

And such a conceptual analysis is one of the main distinguishing features of this book by Vikram Athalye and Emmanuel Haven, in contrast to the majority of previous monographs on quantum-like modelling. Vikram Athalye has a keen interest in the foundations of physics, in general and of quantum theory, in particular (specifically with reference to the quantum-to-classical transition). He was one of the pioneers in investigating the Leggett–Garg inequality (LGI) using precessing nuclear spins. By applying the methods of quantum information processing to investigate quantum foundations, his work (along with S. S. Roy and T. S. Mahesh) gave a theoretical protocol to test LGI as well as resulted in an experimental exploration of the relation between macro-realism and decoherence. Emmanuel Haven was the pioneer in applications of Bohmian mechanics to finance. This approach to quantum finance was developed theoretically (in the works of Haven, Choustova, and the author of this foreword) and tested experimentally on sets of real financial data. Thus, together, the authors represent an excellent team having both a critical understanding of the foundations of quantum physics and of quantum information processing, as well as substantial experience in applying the quantum formalism to the social sciences.

The authors start slowly with a historical review of physics and the economic sciences and introduce the basics of the classical and quantum theories, so any newcomer can proceed further. Then they highlight the information dimension of quantum theory. This quantum-information approach is applied to economics and finance, including the important and intriguing debate of asset prices and information. The quantum information theory provides the possibility to structure this debate within quantum theory of measurements. With reference to such developments, the authors take a review of certain models representing the quantum-like research.

As already mentioned, the absence of the canonical quantization procedure in the social scientific setting makes quantum-like representation of the observables challenging in the social sciences. In quantum physics, besides the quantization procedure, one can also relate the classical and quantum quantities via the Bohmian approach to quantum theory. This theory is characterized by introduction of an additional potential energy – the so-called quantum potential energy. One of the main features of Bohmian quantum mechanics is that it is an inherently nonlocal formulation – the quantum potential energy is nonlocal. It is interesting to note here that David Bohm and Basil Hiley applied Bohmian mechanics to modelling cognition and consciousness. In such models, the quantum potential energy is regarded as an information potential energy.

The authors elegantly and systematically deal with the challenge of representing observables in the social context in quantum-like manner by paying serious attention to the physical foundations and the interpretations of quantum theory. One of the novel features of the discussion presented in the book is that the authors also take into account philosophical insights from the theories of analogical arguments and reasoning, and come up with a strategy for constructing quantum physics–analogous models for the economic sciences.

Without restricting themselves only to the Bohmian approach to finance, at the end, the authors apply both the 'pilot-wave' as well as the 'ensemble' interpretation of the wave function to the development of quantum physics-analogous financial models.

The book can be considered as a rare example of an account that not only reviews the application of the mathematical formalism of quantum theory but also extends and generalizes the corresponding approach to a quantum physics-analogous one by bringing together research areas as diverse as foundations of physics, philosophy of science, and finance. It thus offers a unique possibility for the reader to be introduced to the essentially multidisciplinary aspects of the 'quantum economic sciences' project via conceptually clear and structured presentation.

Generally, development of quantum models in economics and finance is a project of great complexity and this book is an important step toward its realization.

Andrei Khrennikov, PhD

Preface

At the time of writing this preface to the book, we have seen recent advances in artificial intelligence which probably are at a level of progress never seen before in humanity. The development of the quantum computer may possibly be more imminent than one thinks. We can almost surely claim that, one day, the quantum-logic gated computer will be part of the human experience. What will be the consequences? If for a moment we pause our thoughts and ask ourselves the following question: were we to look back to humanity after having reached those important milestones of spectacular advances in AI and the existence of a genuine quantum computer, where would we place, then, research in the social sciences? What would the research look like? Would the quantum formalism be an integral part of any modelling development in, say, economics or finance? Or would it not be? Would social science at that futuristic time then just be embedded with those formalisms? Or rather, would (quantum) physics be mapped (albeit in a justificatory way) onto social science? Clearly, we have no answers. Although it may well be perceived as wishful thinking that there would be a justificatory mapping between (quantum) physics and social science, this book's agenda is precisely to start to engage with that thinking.

In this work we start from first principles, surveying some of the basics from both classical and quantum physics, and then inform the reader of some of the twentieth-century-based key developments occurring in economics and finance.

The first big moment in the book comes about when we survey some of the key work which has been performed in applying the formalism of quantum mechanics to, first, decision-making, and in a subsequent chapter, to finance. Decision-making formalisms are an integral part of most economics/finance models, with the exception of a very important area of finance, which is derivative pricing.

The second big moment of the book, which really makes this work different from many books in the area, is that we then start arguing, as precisely as possible, why a mapping between quantum physics and social science is feasible. We emphasize the all-important concept of an ensemble, from both classical and quantum physics perspectives, and then map that notion into social science. We then move onto another big step: how can we compare electrons with (decision-making) agents? Can the double-slit experiment, so basic in quantum physics, lead to our proposal to consider two-preference interference via a careful specification of a set of classical (linear) potential energy functions?

We hope this book informs the reader that maybe we can go beyond the pure mathematical formalism of quantum theory and reach the mapping stage. In other words, can we go beyond the great achievements of quantum formalism to social science? Can we move on to a carefully argued mapping between quantum physics and social science? As usual with replies to those sorts of questions, one employs the usual words – time will tell – and also – the models will need testing – and so on. There is nothing untrue about those usual words at all, but the book here indeed is not at this stage. It merely provides another little stone, thrown into the lake of uncertain knowledge: shall the ensuing ripples be informative or not? We invite the reader to decide.

Acknowledgements

- All versions of this work may contain content reproduced under license from third parties. Permission to reproduce this third-party content must be obtained from these third parties directly.
- Bartha, Paul. 2022. Analogy and analogical reasoning. *The Stanford Encyclopedia of Philosophy*. (Summer 2022 Edition), Edward N. Zalta (ed.), https://plato.stanford.edu/archives/sum2022/entries/reasoning-analogy/. The Stanford Encyclopedia of Philosophy (SEP) holds an exclusive license, granted by the author of the cited entry, to distribute its content in electronic and digital formats, and the editors have granted us permission to reproduce and distribute an extended excerpt from this entry in that format.
- von Bloh, K. and Athalye, V. 2022. The Double-Slit Experiment in a Uniform Potential. *Wolfram Demonstrations Project*. https://demonstrations.wolfram.com/TheDoubleSlitExperimentInAUniformPotential/ Wolfram Research, Inc., Mathematica, Version 13.1.0, Champaign, IL (2022). www.wolfram.com/mathematica/.

Part I

Introduction and Background

1

Underpinnings of Interdisciplinary Inquiry

In this chapter, we give a brief, and very nonlinear, historical tour of both physics and the economic sciences. We start with Thales of Miletus, the father of science, and provide an overview of an almost twenty-six-century-old quest for the universal laws governing 'happenings' within reality and hence, for an understanding of its nature. Then, we move into some of the essentials of economics and finance. We discuss ideas behind the choice of agents, the grand types of agents and markets; and discuss the role that Kolmogorovian probability plays in economics. This book will be *preoccupied* with a specific generalization of (Kolmogorovian) probability, quantum probability, which will be covered in later chapters of the book. In the last section of this chapter, we provide abstracts of the various upcoming chapters.

Parts II and III of the book deliberate respectively about the quantum-like and the quantum physics–like approaches to modelling some of the activities and processes studied under the economic sciences. With hindsight, the common ground for both these approaches can be seen as a certain kind of hypothesized similarity between the ways in which the natural and the social phenomena are described and analyzed. Briefly speaking, the *quantum-like* approach tacitly assumes the usefulness of the mathematical language and the related methodologies of (quantum) physics analogously in the economic sciences, whereas the *quantum physics–like* approach is based on the conjecture that the analogy – between the happenings that are conventionally studied under physics (where the participating systems are physical systems) and those studied under the economic sciences (where the participating systems are social systems involving 'agents') – exists with regards not just to the mathematical forms but also to the interpretations of the quantities in terms of which natural phenomena are modelled in the physical theories.[1]

[1] We elaborate on this point in Chapter 8.

Before getting into the foundations and the intricacies of these two modelling approaches that assert connections between the apparently distinct and independent domains of physics (to be understood as the 'source' domain) and the economic sciences (to be understood as the 'target' domain), we think it is important to keep in mind how physics can be seen as an ever-evolving field of knowledge about the nature. Also, it will be essential to have a certain idea of the salient features of the structure of physical theories.

So let us begin with the question: what is physics about?

Physics is generally understood as the natural science that deals with the properties of matter and energy. Specifically, it studies the behaviour of matter in terms of its 'motion' through space-time, under the action of forces that can have gravitational or electromagnetic or nuclear origins. It makes the workings of the universe comprehensible in terms of physical laws. Physics is concerned with aspects of nature at all levels – on the one hand, it teaches us a great deal about the functioning of the universe at the cosmic, macro-scales and on the other, it offers explanations of phenomena at the subatomic and subnuclear scales.

Physics shows its presence in our day-to-day activities and experiences as well. For example: it explains what causes lightning, rainbows, blue sky, aurora and also gives the reason behind using a metallic can with wood on the outside as an ice-cream maker! It explains how a microwave oven heats and cooks food and tells why you should be more careful while eating a hot pizza than drinking hot soup sip by sip at the same temperature. It explains what makes an acoustically proper auditorium and what distinguishes an electric guitar from an acoustic guitar and so on. To explore many more interesting daily-life examples showing the presence and usefulness of physics, see for example [1].

Physics is versatile: it does not just explain fundamental natural phenomena but is also practically useful in many other natural as well as applied scientific disciplines such as chemistry, biology, palaeontology, climatology, geology, and medicine. Also, it is the basis of all engineering and technology.

The meaning of physics has evolved from the ancient era, in which it was understood as natural philosophy, to the modern times, where it now represents not just natural philosophy but also the most fundamental, all-inclusive and hence, universal science (see [2]). These facts about physics lead to an intriguing query: how did physics, which originally represented the *philosophy* of nature, evolve into its modern phase of the philosophical as well as the *scientific* knowledge of nature? The answer to this query lies in the longstanding history of the developments of physics [3]. We, therefore, open our discussion (Section 1.1) with a brief review of historical developments in physics. Then we present our viewpoint on the generic structure of the physical theories of motion (Section 1.2). In Section 1.3, we discuss some of the main objectives of the economic sciences. Section 1.4 gives abstracts of the remaining chapters.

1.1 Physics: From a Brief History to the Economic Sciences

It is worth having a look at some milestones from the history in order to understand how physics is an ever evolving field of fundamental, philosophical, and scientific knowledge about nature, and this way, how its all-inclusive ontology could be seen to provide room not just for the phenomena and processes studied under other sciences and technology, but also for the agent-based activities modelled in the social sciences. We think such an understanding is essential for the development of physics-*analogous* models for the economic sciences.

Of course, a detailed account of the historic developments in physics is beyond the scope of this book and can be found elsewhere, see, for example, [3] and [4]. Here we restrict ourselves to only a sketch of the history of physics. We chronologically follow the four phases of the developments: ancient times (1.1.1), the scientific revolution (1.1.2), the birth of modern physics (1.1.3) and the modern version of the quest for the nature of reality (1.1.4).

1.1.1 Ancient Times

Historical records show that the movement for a rational understanding of nature started around the seventh century before the Common Era (BCE) in ancient Greece. At that time the most prominent of the natural philosophers was Thales (624–565 BCE) from the city of Miletus, who was dubbed the father of science. His forecast of the solar eclipse, which was to be visible from Miletus at that time, promoted the importance of systematic observation of nature among the Greeks. Thales declared that every natural process had a *cause* and *change* is the fundamental and essential part of nature.

Heraclitus of Ephesus (540–475 BCE) held similar views about nature. He emphasized that things perceived by the senses are constantly in a state of becoming and that change is the only reality. This is the so-called *doctrine of becoming*. The idea of reality for Democritus (470–400 BCE), however, was in sharp contrast to Heraclitus'. According to Democritus, reality was made up of two primary components: concrete solid structures (similar to the modern concept of atoms) and empty space, such that these primary components are unchanging and fixed. This is the so-called *doctrine of being*. Although the idea of Democritus on the fundamental nature of reality appears to be 'in sync' with the modern scientific *doctrine of atomism* (i.e. indestructibility of matter, see Subsection 1.1.2), in ancient times it was put forth more as a hypothesis rather than an experimental finding.

Plato (427–347 BCE) was familiar with the doctrine of becoming. He had a great admiration for the exactness of mathematical principles. This exactness, according to him, was deeply connected to the *eternally existing, perfect ideas* – that is, the idealized mathematical forms – abstracted from the physical world (see [5]).

He asserted that such ideas only are precisely definable and must be differentiated from the actual things from the physical world, perceived by our senses, for those things are never the same but *ever-changing components* of the reality. Aristotle (384–322 BCE), who was the pupil of Plato, pointed out that this process of abstraction was necessarily erroneous if used to describe the physical world, for it leaves out those conceptions in which lies the true, inherent nature of matter – the nature of constant becoming.

Aristotle developed the notion of *space* as an absolute, fixed background in which the *motion* of the things takes place [4]. He considered space as similar to a nonportable vessel and as a whole, motionless and fixed. This Aristotelian idea of space can be briefly explained in terms of the geometric representation due to Euclid (330–260 BCE). If we assume Euclidean 3-space (denoted by $\mathbb{E}^3$) as representing the geometry of the Aristotelian physical space, then Aristotle's notion of space means that the points belonging to $\mathbb{E}^3$ continue to maintain their identity from one instant of time to the next. Aristotle understood *time* as an attribute of movement, independent of the observer. Also, for Aristotle, the flow of time had an appearance that is similar to the motion of a body along its trajectory, such that the 'now' (i.e. the instant representing the 'present') is like the body in motion, separating the 'past' from the 'future'. To summarize, the Aristotelian ideas imply *absoluteness* of space and time (see [5] sec. 17.1).

Aristotle also pioneered a qualitative viewpoint on the causality underlying the occurrence of the natural processes. The Aristotelian idea of causality is expressed in terms of the doctrine of four kinds of causes: material, formal, efficient, and final. These are to be understood as distinct (though, interrelated) categories for posing causal queries pertaining to the behaviour of natural things or to the occurrences of natural processes and it can be briefly explained as follows. The material cause speaks for what undergoes change in a given process, namely the material involved in the process. The formal cause is about what form (e.g. shape) of the material will come to exist in the process. The efficient cause indicates the external influences/interventions behind the happening of the process or the principle(s) due to which the process takes place. The final cause points to the aim to achieve a certain end result for the sake of which the entire process (start to end) takes place.

1.1.2 The Scientific Revolution

Let us now jump from the 'BCE' era to the period between the fifteenth and the nineteenth centuries, which witnessed the 'scientific revolution'. At the beginning of the previous section we mentioned the forecast of the total solar eclipse by Thales of Miletus as marking the origin of scientific thinking. The revolution in scientific thinking also originated in the context of astronomical observations, related to the planetary motions. The beginning of modern astronomy is marked by the detailed reasoning provided by Nicholas Copernicus (1473–1543) on the heliocentric model.[2] Based on this work and on the accurate observations made by Tycho Brahe (1546–1601 CE), Johannes Kepler (1571–1630 CE) formulated the basic laws of planetary motion (see [4, 5]).

It was Galileo Galilei (1564–1642), who discovered that the physics of the state of rest and the physics of the state of uniform motion are indistinguishable. This fact leads to the so-called Galilean principle of relativity, which can be stated as follows: the laws of motion must be the same in all coordinate systems moving with uniform velocities with respect to each other. The principle implies that the mathematical equations expressing the laws of motion must have the same form in all coordinate systems moving with uniform velocity relative to each other. In such coordinate systems – which are also called the inertial frames of reference – Newton's first law (the law of inertia) holds. The Galilean principle of relativity replaced the Aristotelian idea of absolute space. It implied a space-time picture, in which, the $\mathbb{E}^3$ is not fixed in time, rather, the entire $\mathbb{E}^3$, representing the *arena* for the dynamics of the physical world, evolves with time. Consequently, in the Galilean scheme, it is meaningless to say that a given point belonging to $\mathbb{E}^3$ is the same as it was a minute ago! It is said that the idea of the physical equivalence between the state of rest and of uniform motion led to Galileo's acceptance of the motion of the Earth through space without us noticing it and it led to his support for the Copernican heliocentric model. This way Galileo could note that pure philosophical ideas about the nature of reality may lead to erroneous conclusions, which could be avoided by developing *natural philosophy* in the light of *mathematical analysis* (see [5], section 17.2).

René Descartes (1596–1650 CE) sought to express natural philosophy by considering all perceived phenomena as reducible to the motion of an invisible sea of particles. He saw a similarity between the motion of the planets and that of the terrestrial objects. He pointed out that these two apparently different kinds of motion must obey universal principles. His famous *Cartesian coordinate system* allowed the motion of a particle to be expressed as the plot of a curve – known as the *trajectory* of the particle – corresponding to a precise relation between the variables representing the coordinate *axes*. This was the beginning of the scientific revolution in sixteenth-century Europe – rooted in the realization that the motion of bodies – whether celestial or terrestrial – can be *universally* characterized through *mathematical* descriptions (see [3]).

It was Sir Isaac Newton, who, for the first time in history, formulated this anticipated universality in terms of three laws of motion along with the law of universal

[2] The heliocentric model represents the Sun as the centre around which the planetary motion takes place. It is the accepted astronomical model of the Solar System.

gravitation. In 1687, he presented this work in detail and with *mathematical rigour* in his famous book the *Principia* (short for *Mathematical Principles of Natural Philosophy*). Later on, it turned out that the mathematical equations of the Newtonian dynamical scheme retain their forms in all the inertial frames of reference and hence, are consistent with the Galilean principle of relativity (see [5] section 17.3). During the eighteenth century, the Newtonian mechanics were further developed by several scientists using analytical methods of calculus. The resulting formulation is known as *classical mechanics*.

The nineteenth century witnessed profound developments in physics along two directions: One gave rise to the classical theory of *electromagnetism* and the other to the scientific formulation of *thermodynamics*. The latter laid the foundations of a new approach to the scientific understanding of nature, which later on became known as *statistical mechanics* (see [4] chapters 9 and 10).

In the 1860s James Clerk Maxwell published his work on the unification of electric and magnetic fields. Built on previous work by several scientists (including Faraday, Gauss, Ampère, and Weber), he provided a set of coupled partial differential equations, which described the generation of electric and magnetic fields due to charges and currents. Maxwell's work supported the wave description of light and established that light is an electromagnetic phenomenon. Along with Lorentz's force law, Maxwell's equations formed the foundation of classical electromagnetism.

In the 1840s, the investigations of Mayer and Joule established the interchangeability of motion and heat (the equivalence between mechanical energy and heat) and this led to the theory of energy conservation. Understanding such a connection proved to be important for the development of steam engines based on the earlier work of Sadi Carnot published in 1824. Carnot captured some of the ideas of thermodynamics, for example, in terms of the efficiency of an idealized engine. Such ideas provided a basis for the formulation of the first law of thermodynamics, which is a restatement of the law of conservation of energy. During the 1850s and 1860s Lord Kelvin and Clausius analyzed the fact that heat does not spontaneously flow from a colder body to a hotter. This analysis led to their statement of the second law of thermodynamics. The principles of thermodynamics seemed to imply that the flow of energy between matter and a radiation field – with both themselves assumed as *unbroken continua* – was continuous (see [6]).

However, the developments in the second half of the nineteenth century showed a revival of the *doctrine of atomism*. To the atomists, matter was not continuous, rather it was made up of discrete structures such as atoms and molecules. Moreover, the theories of mechanical motion could be applied to atoms and molecules to calculate the thermodynamical properties of material substances. For example, in 1860, Maxwell discovered the *statistical distribution law* (also known as the

Maxwell–Boltzmann probability distribution) for the velocities of the molecules of a gas at a given temperature (see [4] section 9.2).

1.1.3 The Birth of Modern Physics

At the end of the nineteenth century, one of the unsolved problems was understanding the observed behaviour of cavity radiation. Its theoretical explanation required a distribution function for the intensity (or equivalently, energy) of electromagnetic radiation, at a given temperature, across a range of radiation wavelengths. The derivation of the correct formula for the function – provided by Max Planck in 1900 through the introduction of his famous constant, the 'quantum of action', denoted by 'h' – required the assumption that the electromagnetic energy is absorbed and emitted by material substances in *discrete* packets[3] (*quanta*, also called *photons*) rather than in a continuous manner. In 1905, Albert Einstein used Planck's 'quantum' hypothesis to successfully explain the experimentally observed properties of the photoelectric effect. Planck's derivation marked the beginning of the most profound revolution, the *Quantum Revolution*, in mankind's scientific understanding of the nature of reality made up of matter and radiation; and the nature of 'motion' (see [6]).

The year 1905 witnessed yet another profound revolution in our understanding of the nature of space-time and its relationship with physical observations, due to Einstein's 'special theory of relativity' (STR), which replaced the Galilean idea of space-time and principle of relativity. The origin of the STR lies in the fact that the speed of light appears as a constant in one of the Maxwell's equations of electrodynamics. Also, as exemplified by Faraday's law, the laws of physics must be the same for all the observers. These two inputs served as the postulates for Einstein's STR. With the speed of light in vacuum as constant and to be taken as the upper limit for the speed of anything in motion, according to the STR, the simultaneity of two events is not an absolute idea, rather, it is observer-dependent,[4] in the sense that, two events, appearing to happen simultaneously for an observer, say O, do not appear as simultaneous for the other observers having relative motions with O.

Moreover, in terms of the STR, it turns out that not just the space (lengths and distances), but also the time (i.e. temporal interval between the occurrence of two events) is a relative idea and, thus, the STR provided an understanding of the geometric structure of space-time that is more general than the Aristotelian ideas of the absoluteness of space and time, and also, than the Galilean ideas of relative space

[3] Having energy $E = h\nu$, with ν being the frequency of the electromagnetic radiation.

[4] Here, by an observer we mean an inertial reference frame represented, in general, by a $(3 + 1)$-dimensional coordinate system, with 3 spatial and 1 temporal dimensions.

and absolute time (see [4] chapter 12).[5] Also, the STR established the equivalence of mass and energy as expressed by Einstein's famous equation, $E = mc^2$.

In the meantime, with the 1897 discovery of the electron by J. J. Thomson, the ancient notion of an atom as an indivisible elementary entity evolved to the idea that an atom is a system having an internal structure. This discovery stimulated an interest in the investigation of atomic structure. In this context, Rutherford's alpha scattering experiments (which he performed with his research associates Geiger and Marsden during 1909–11) represented pioneering work. These experiments established that most of the atomic mass is concentrated in a small central region, called the (positively charged) nucleus and it is surrounded by a system of (negatively charged) electrons. However, such a model was considered inherently unstable in terms of the classical electromagnetic theory. In 1913, based on the 'quantum' idea of Planck, Niels Bohr improved Rutherford's atomic model. The Bohr atom model gave a detailed theoretical explanation for the observed spectroscopy of the hydrogen atom (see [6], chapter 3). In 1923, Arthur Compton and Peter Debye showed how the light quanta (photons) exhibit a particle-nature by demonstrating their scattering from electrons.

In spite of this progress in our understanding of the fundamental nature of reality, a proper framework – based on a new set of principles – was required which could reconcile the *wave* and the *particle* aspects of light into a single theory that would account for the stable internal (discrete energy-level) structure of the atoms as well. In this situation, a major breakthrough came with Louis de Broglie's 1924 hypothesis of 'matter waves', according to which, the nature of matter, just like radiation, is *dual*, exhibiting wave-like as well as particle-like behaviour. Building on de Broglie's approach, Werner Heisenberg, Max Born, and Pascual Jordan developed the matrix formulation (1925) and Erwin Schrödinger developed the wave formulation (1926) of modern quantum theory. These developments were followed by Heisenberg's formulation of the uncertainty principle (see [4, 6]).

1.1.4 Quantum Theory and the Nature of Reality

To make sense out of all these developments and to understand the implications of quantum theory with regard to the nature of reality, its interpretation – beyond the mere mathematical (wave and matrix) formulations – was required. One of the earliest attempts made to interpret quantum theory was that which grew out of the interactions of Niels Bohr and Werner Heisenberg during 1925–27 held in

[5] In 1915, Einstein generalized the STR to include gravity. The resulting theory became known as the general theory of relativity (GTR). In the GTR, Einstein described gravity in terms of the *curvature* of space-time arising due to the presence of matter. This description replaced the Newtonian idea of 'gravity as a force.'

Copenhagen. The ideas of complementarity and of wave function reduction (see [6]) formed the basis of this so-called Copenhagen interpretation.[6]

As it turns out, quantum theory allows a variety of approaches to interpreting the nature of reality and the behaviour of natural things. Even though these approaches differ from each other (for instance, some of the approaches give importance to an ontological meaning, whereas some others put an emphasis on a pure epistemological understanding of the theory), they all agree on the predictions and the experimental "end results" of the theory.[7] In the 1920s, when it came to interpreting the "meaning" of quantum theory, perhaps the strongest and the foremost critic of the theory was Albert Einstein!

Einstein was firmly committed to a worldview based on three tenets: local reality, causality, and determinism. By anticipating a harm that could be potentially caused by quantum theory to these tenets, in the fifth Solvay conference, held at Brussels in October 1927, Einstein expressed his reservations about the modern formulation of quantum theory and also, about the nature of reality implied by the Copenhagen interpretation of quantum theory [6]. In a paper published in 1935, he and his coworkers, Boris Podolsky and Nathan Rosen, put those reservations more clearly (see [7]).[8] By analyzing a thought experiment, they claimed that the quantum mechanical description of physical reality is incomplete. They also expressed their belief that a more complete, "causal and deterministic" theory was somehow possible. It was anticipated that any such theory would necessarily invoke new variables, though *hidden* from observation. In the meantime, in his book *Mathematical Foundations of Quantum Theory*, John von Neumann provided a mathematical proof (based on certain assumptions) that hidden variable theories are impossible.

Despite the Copenhagen interpretation of quantum theory and the impossibility proof by von Neumann, the efforts to interpret it in alternative ways and to develop its (potentially) more complete version were continued. A major breakthrough came in 1951 with a causal formulation and interpretation of quantum theory, which was proposed by David Bohm following the earlier efforts by de Broglie. He asserted the impossibility of having a theory of mechanically determined hidden variables that can reproduce the results of the quantum theory [8]. Based on this assertion, Bell, by means of the Bohm–Aharonov version of the EPR experiment (1957), proved an inequality and established that the results predicted by quantum theory can violate the inequality. He concluded that it is possible to find measurement configurations which would demonstrate that the quantum phenomena are incompatible with any *local* hidden variable theory and hence, local reality.

[6] The wave-like and particle-like behaviours of the same elementary microscopic physical system, such as an electron, are complementary to each other, rather than contradictory.

[7] We will briefly discuss some of the "quantum interpretations" in Subsection 8.1.3.

[8] In the same year, Bohr published a reply to the EPR (Einstein–Podolsky–Rosen) argument.

The 'Bell inequality' opened doors to a whole lot of several intriguing theoretical and experimental investigations on the fundamental 'quantum' nature of reality. In 1969 John Clauser and his coworkers came up with a proposal for testing Bell's inequality experimentally [9]. In the early 1980s the inequality was observed to be violated in the comprehensive experiments performed by Alain Aspect and his colleagues. A Bell-inequality violating 'quantum' state (which is also called simply a Bell state or an entangled state) can be prepared and among other things, be used, for example, for quantum teleportation, which is a technique by means of which quantum information can be transferred over distances. Such a technique was proposed for the first time in 1993 by Charles Bennett and his co-workers and was experimentally realized for the first time by Anton Zeilinger's group. This way, starting 26 centuries back, from the ancient history of physics, we arrive at the 2022 Nobel prize in physics: it was awarded to Clauser, Aspect, and Zeilinger "for experiments with entangled photons, establishing the violation of Bell inequalities and pioneering quantum information science."

Thus, no doubt, the quest for a deeper and deeper understanding of the nature of reality and of the workings of the universe – at all levels, from the 'micro' to the 'macro' – is still on. And, what we discussed earlier is just a very brief account of this centuries-old quest. Nevertheless, we hope that it sheds light on the ongoing efforts of physicists and the natural philosophers to provide a more and more accurate description and comprehension of reality, through models and experiments. By keeping in mind such efforts, we can surely consider 'physics' as the ever-expanding system of our philosophical *as well as* scientific knowledge of the fundamental natural processes taking place within the universe.

1.1.5 The Universe includes Economic–Financial Activities As Well

All these developments indicate that, as an academic discipline, physics is often perceived as the natural science restricted to the explanation of the phenomena and processes involving *inanimate* things (e.g. electrons, atoms, automobiles, planets, and stars). Yet, neither its original Latin (*physika*) and Greek (*phusika*) meanings ('natural things' following from the term *physis* meaning nature) nor the ever-expanding system of fundamental natural philosophy and natural science that it represents imply such a restriction on the domain of applicability of physics; see [10].

In fact, owing to their inherently universal and all-inclusive nature, physical laws – whether already known or yet to be known – must be applicable in principle to anything and everything happening in the universe. And, in this way, physics can be seen as providing ontological roots for the other systems of knowledge, including – as already mentioned at the beginning of this section – the life sciences.

Indeed, the previous century and also the current century so far have witnessed the influence of the developments in physics on the subsequent developments in the life sciences (see, e.g., [11]) including the investigations on the physical origins of the human consciousness (see, e.g., [12]). If viewed in this manner, the 'physics-ontology' could be seen as embedding the ontology for the social sciences as well, for after all, these sciences involve modelling and analyzing the behaviour of systems whose building blocks are nothing but the conscious human beings that are very much the parts of the universe!

Having said this, we admit that developing a first principles, physics-ontological model of the working of an individual conscious human brain is itself extremely challenging and therefore, the construction of even simple physics-ontological dynamical models of the social activities involving a large number of human brains as participants is quite far from being simple.[9]

Thus, on the one hand, the ontology of the social sciences can be seen, in principle, as a part of the (universal, all-inclusive) ontology of physics; on the other hand, a physics-ontological stance for modelling social phenomena would not take us anywhere. Nevertheless, when perceived from a dynamical viewpoint, the happenings modelled in the social sciences (for instance, those in economics and finance) seem to bear certain similarities with those modelled in physics (as discussed in Chapter 8), and that is why already several modelling approaches from theoretical physics have been applied in the social sciences.

Throughout the book, our discussions aim to address the following query: how could the rich variety of the dynamical formulations, ideas, and notions from physics including the variety of interpretations of reality provided by physicists be used in economics and finance? In Chapter 8, this query is addressed by proposing a detailed strategy that could lead to the construction of models for economics and finance in a physics-*analogous* – rather than in a physics-*ontological* – manner. Part III – to a large extent – will be preoccupied with the application of such a strategy to describe and analyze the stochastic evolution of the asset price.

As prerequisites for Parts II and III, it is essential to know both the generic structure of the physical theories of motion (Section 1.2) and the key ingredients of a generic economics/finance model (Section 1.3).

1.2 Generic Structure of the Physical Theories of Motion

Two of the most prominent features of the physical theories of motion developed during and after the so-called scientific revolution were the *mechanistic*

[9] A physics-ontological model could be interpreted as a model constructed with a reductionist's approach. As per its dictionary meaning, 'reductionism' refers to a way of analyzing and describing a complex phenomenon in terms of its simple or fundamental constituents.

philosophical basis (for stating the dynamical laws of the theory of motion in question) and *mathematization* (i.e., the expression of those dynamical laws in precise mathematical language). The mechanistic philosophical basis gave rise to an understanding of a given dynamical process in terms of physical quantities such as forces, fields, and energies, whereas the mathematization required describing the relationships involving those physical quantities in mathematical terms and expressing the changes in those quantities with respect to space and/or time by means of differential equations.

Today, with hindsight, we might say that any physical theory of motion – for example classical mechanics, statistical mechanics, electrodynamics, quantum mechanics and so on – represents a certain *dynamical paradigm* and assumes a certain nature of *space-time* ([13], [5]). These aspects can be considered as the main components of the framework of any physical theory of motion and hence, deserve attention when it comes to developing a strategy for constructing *physics-analogous* models of the economics-financial reality. In the following we elaborate on both these aspects.

A physics paradigm can be said to represent a generic pattern of a variety of similarly occurring natural processes. Specifically, it refers to a (dynamical) law or a set of laws of nature governing an occurrence. As already remarked, these laws are stated in terms of relationships between well-defined physical quantities. Specifically, a dynamical law accounts for the time-evolution of the states (or of the probability of the states) of the participating physical systems. Examples of the physics paradigms include Newtonian physics (NP), special relativity (SR), quantum mechanics (QM), quantum field theory (QFT), and general relativity (GR).[10]

Physical quantities involved in the description (kinematics) and also in the causal analysis (dynamics) of the motion require representations with respect to a well-defined space-time structure – providing the 'arena' for the natural processes to happen. But, is the idea of space-time similar to that of an arena? Let's consider the colloquial meaning of the term 'arena'. An arena, also known as the 'stage', is a place or an area of some activity, for example, a drama or an opera or a concert or other live performances, where the performing artists can move in a manner that can be resolved in terms of the left-right, to-and-fro, and up-and-down components. As a viewer we can also notice that all the pieces of the entire performance – dialogues or dances or musical renditions – do not appear 'all at once' within the arena. Rather their appearances follow a certain chronology. Some pieces of the performance appear 'before' and some appear 'after'.

Just like we need some arena for a performance (of artists) to take place, there has to be some arena for the natural processes to take place within the universe.

[10] See, e.g., [13]. In the present book, for our physics-inspired explorations in the economic sciences, we will be concerned with the paradigms of NP and QM only.

In physics, such an arena is provided by the geometric structure of the so-called *space-time* – a combination of the terms: 'space' and 'time'. Similar to an arena, the physical 'space' provides a three-dimensional scope for the (left-right, to-fro, and up-down) movement of the things in nature, whereas the 'time' parameter provides a chronology-label by preventing the interconnected parts of a whole natural process from occurring all at once.

Of course, this does not mean that the idea of an arena and that of a space-time are exactly similar to each other, for the topic of the geometry of space-time is much more intricate than the simple idea of an arena. For instance, there is a variety of space-time geometries, for example, Aristotelian, Galilean, Newtonian, and Minkowskian (see [5]). Also, different models of the space-time structure are suitable for different paradigms. For example, the dynamics under the 'NP' paradigm can be suitably represented on the fixed Galilean space-time (see the discussion that follows) and hence, is understood as a background-dependent paradigm. The paradigm of SR is also background dependent. For the dynamics of the QM paradigm, for all the practical purposes, a fixed Galilean space-time background is assumed, whereas general relativity is an example of a background-independent paradigm that models the space-time structure itself as 'dynamic'. Moreover, the issue is not always about distinguishing between a fixed background space-time and a dynamic one. According to some physicists, the dynamics of quantum theory are applicable to space-time as well and therefore, space-time should be considered as an emergent structure having classical appearance. However, we will not worry about such intricacies and will rather be concerned with only the Galilean ideas of space and time.

Thus, with reference to the preceding discussion, we can think of a physical theory of motion as a combination of a certain *paradigm* and a well-defined space-time structure, that is, an *arena*. To explain briefly how a typical theory of motion could be seen to involve a paradigm-arena combination, we present the analysis of a familiar situation (an object – modelled as a 'particle' – falling down due to gravity on the Earth's surface) in the following subsections.

1.2.1 Motion of a Particle under Gravity

Consider an object falling down from a height toward the centre of the Earth. The dynamics of such a motion (the *cause*) can be analyzed in terms of the paradigm of NP. According to the NP paradigm, gravity is interpreted as an attractive force acting between any two objects. It arises from the gravitational field existing in the vicinity of those objects on account of their masses.[11] According to the law of universal gravitation, the gravitational force (F_g) between two objects is –

[11] We note that the paradigm of general relativity provides a more detailed description of gravity and a more accurate account of gravitational processes than that provided by the paradigm of NP.

- directly proportional to the product of their masses (say, m_1 and m_2), and
- inversely proportional to the square of the distance (r) between their centres.

In terms of these symbols, the mathematical relation expressing the law of universal gravitation is

$$F_g = G\frac{m_1 m_2}{r^2}. \tag{1.1}$$

Here, $G \approx 6 \cdot 6 \times 10^{-11}\ \mathrm{m^3 \cdot kg^{-1} \cdot s^{-2}}$ is known as the universal gravitational constant.

In particular, the relation (1.1) also gives the gravitational force due to the Earth experienced by the particle: Let m and M denote the masses of the particle and the Earth, respectively and r be the radius of the Earth. If h is the height of the particle above the Earth's surface, then its distance from the Earth's centre is $(r + h)$. For small heights, $(r + h) \approx r$ and Equation (1.1) can be rewritten as:

$$F_g = G\frac{Mm}{r^2}. \tag{1.2}$$

1.2.2 Gravity and Newton's Second Law of Motion

The motion of the particle under gravity can be explained by applying Newton's second law, which, briefly speaking, relates the acceleration of a particle with the resultant force acting on it, divided by its mass:

$$\mathbf{a} = \frac{d\mathbf{v}}{dt} = \frac{\sum \mathbf{F}}{m}. \tag{1.3}$$

In the preceding relation $\mathbf{a}$ is the acceleration vector, $\mathbf{v}$ is the velocity vector and the resultant force $\sum \mathbf{F}$ is obtained by taking the vector sum over all the forces acting on the particle.

A comparison of Equation (1.2) with Equation (1.3) implies that the gravitational force exerted by the Earth on a particle *causes* an accelerated motion of the particle toward the centre of the Earth. The magnitude of the so-called acceleration due to gravity is denoted by g and is given by $\frac{GM}{r^2}$. Its approximate value is $9 \cdot 8\ \mathrm{m/s^2}$ with respect to the Earth. Thus, we can rewrite Equation (1.2) as $F_g = mg$.

Due to the gravitational force, the speed of the falling particle goes on increasing steadily by (approximately) 9.8 m/s every second.[12] Its momentum, being the product of its mass and the velocity ($\mathbf{p} = m\mathbf{v}$), also goes on increasing at a steady rate. This rate of change in momentum is defined as the force acting on the particle.

[12] Of course, here we ignore the effect of the upward force on the falling particle due to the buoyancy of air. This is a velocity-dependent force, which increases with the increase in the particle velocity. Therefore, more realistically speaking, when such a force goes on increasing and becomes equal to the force due to gravity, the velocity of the particle becomes constant and is called the 'terminal velocity'. See e.g. [14] for details.

Thus, in terms of the linear momentum of the particle, Newton's second law can be reexpressed as

$$\mathbf{F} = \frac{d\mathbf{p}}{dt}. \tag{1.4}$$

1.2.3 Kinematic Considerations

We can now turn to the kinematics (the *description*) of the particle's motion. It is given in terms of different kinematic quantities, such as time, position, velocity, and (constant) acceleration. Within the paradigm of the Newtonian physics, there is a set of kinematical equations (consistent with Newton's laws of motion) relating these quantities with each other. For example, for the particle's motion under gravity, its initial [$v(0)$] and final [$v(t)$] speeds (in one dimension) are related by

$$v(t) = v(0) + gt \tag{1.5}$$

and the changing position as a function of time [$x(t)$] is given by:

$$x(t) = v(0)t + \frac{1}{2}gt^2. \tag{1.6}$$

As expressed by Equation (1.3) the (instantaneous) acceleration of the particle is the time derivative of its (instantaneous) velocity. Similarly, from Equations (1.5) and (1.6), it can be easily seen that the (instantaneous) velocity of the particle is the time rate of change in its (instantaneous) position.

1.2.4 The Galilean Space-time

The discussion from the previous subsections provides some of the important kinematical and dynamical elements of the NP paradigm. Now, what about arena? It can be verified that the relations (1.1)–(1.6) imply (and are consistent with) the following assumptions about the geometric structure of the arena:

1. The 'physical space' is represented by the Euclidean 3-space $\mathbb{E}^3$.
2. The 'time' is represented by the Euclidean 1-space $\mathbb{E}^1$.
3. There is a notion of 'distance', which, for the 'physical space', is the ordinary Euclidean distance: it is the distance between two points defined by assuming that the geometry of the space is Euclidean. Such a distance can be calculated from the Cartesian coordinates of the two points by applying the Pythagorean theorem. In Equation (1.1), 'r' is the Euclidean distance.
4. There is a notion of 'interval', which is nothing but the ordinary time difference between two instants.

Apart from these assumptions, there is also an absolute notion of the temporal simultaneity: this means that different spatial locations (i.e. different points belonging to $\mathbb{E}^3$) can share the same temporal instant (i.e. the same point belonging to $\mathbb{E}^1$). Also, it is not possible to distinguish the physics of 'a particle at rest' from the physics of 'a particle in uniform motion'. The arena based on these assumptions is called the Galilean space-time (or, even better, Galilean–Newtonian space-time). Hence, the relations (1.1)–(1.6) are said to be expressed under the assumption that the fixed background geometric structure of the space-time is Galilean.

The example of the motion of an object under gravity thus provides a plausible explanation of how an inseparable combination of a paradigm (i.e. Newtonian Physics) and an arena (i.e. Galilean Space-time) make up the structure of a physical theory of motion. Along similar lines, we expect that a physics-analogous economics/financial theory should involve a combination of a physics-like paradigm and space-time-like well-defined geometric structure.

1.3 About Economics and Finance

Whilst physics is focussed on explaining natural phenomena, economics and finance will be preoccupied with modelling the economic and financial[13] aspects of society.

At least three key ingredients are needed when attempting to model society as an economy. Markets, being the first ingredient, are the drivers of an economy and the gestation and dynamic expression of those markets is governed via the second key ingredient, namely complex and intricate decision-making processes. We note that markets come in various guises: financial markets such as the stock and commodity markets, but we can easily think of other 'closer to home' markets like the market for sweatshirts or the market for consumable commodities like coffee. Surely, the market for services is another market. An important feature of markets to keep in mind, is that not every market leads to the consumption of the good or service. Futures markets on commodities, as an example, may never lead to the consumption of the underlying asset of the futures contract, but rather the asset is used as a hedging device to fix an (uncertain) future price of (another) to-be-consumed commodity[14].

The third key ingredient of an economy is the economic actor. In economics, this actor can be modelled in various ways, as a representative agent or as an agent who interacts with other agents in a, more or less, prescribed pattern. More on this later.

[13] We shall henceforth use the term 'economics' and explicitly mention the word 'finance' when there is an issue we need discussing which only pertains to finance proper.

[14] A futures security is part of a wider class of securities, called 'derivatives' which also include so called option contracts.

We observe that all three key ingredients listed so far form part of an interacting chain of command, so to speak. As an example, without decision-making, economic actors cannot create markets. Without actors, decision-making becomes irrelevant. The chain of command requires interactions between markets, actors, and decision-making. The interactions are dynamic. We mean by this that actors update their beliefs when making their decisions on the basis of what the outcome is of the market. The price of a commodity or a service is precisely the outcome of the market. Typically we make price to be time dependent.

In what follows, we shall first discuss some of the basic underpinnings of decision-making, followed by how we may view economic actors. Markets are discussed last.

1.3.1 More on Decision-Making

It is intuitive to make the claim that decision-making is performed by many living organisms. There are degrees of sophistication in decision-making in any type of life form, whether they are bacteria, cells, or animals. Living beings must make choices on a continuous time basis. The context of such choice, as we know, varies immensely: for instance, bacteria want to optimize the distance travelled by which they infect a host cell; whilst an 'economic actor' – say, a government – may want to maximize social welfare from running a new social programme. The formalistic structure of economics revolves to a very large extent around the idea that we optimize, whether this is, for instance, social welfare at a macroeconomics level or consumption satisfaction at a microeconomics level.

Let us provide for a simple example. One of the most basic objects in economics is the so-called demand function which establishes a formal relationship between the price paid for a good and the quantity demanded of the said good. The existence of a demand function for a good or service provides for an abstract expression of the existence of the market for the said good or service. Economic decision-making comes into play in this basic formal object in the following way. The downward-sloping demand function is obtained out of a maximization process. What is then being maximized, one might ask? In basic economics, we model the satisfaction we derive out of consuming a good or service with the help of a 'utility function'. In plain words, such as function registers the level of satisfaction one may derive out of consuming for instance, three units of a good as opposed to four units of a good. The maximization of that utility function will be subject to a constraint most of us will face throughout our life times: that is, the budget we need abiding by. Hence, a basic driving principle economics will build upon is, to maximize satisfaction of consumption given the budget we face. For the basic formalism, please consult, for instance, [15] and we shall also attempt to provide for a formal overview on the idea of utility in Section 5.4. Out of this idea came forth a central idea in economics: marginality. For instance, the marginal rate of substitution of a so-called indifference curve (this marginal rate is the slope of that function) will equal the slope of the budget line when we decide to maximize satisfaction subject to the constraint that we will exhaust our available budget against the consumption of the goods making up that level of satisfaction.

We can easily imagine an ordinal ranking of the indifference curves. If both goods are wanted by the consumer, then we could make the simple assumption that 'more is better'. Hence, higher and higher levels of utility will trace out a series of indifference curves with the lower utility level attached to an indifference curve closer to the origin, whilst a higher utility level will occur with an indifference curve further removed from the origin.

1.3.2 More on the Agent: The Rational Representative Agent and Behavioural Approaches

At the root of the maximization problem lies a fundamental trade-off. Resources are finite and time is finite. However, in what we have said so far, there is an underlying key condition we have totally omitted. Although optimizing behaviour, no matter the context, occurs at any stage of life, there is one very important, underlying condition economics will also stress, specifically, at the level of the human decision maker. This is the idea of rationality. The so-called economic agent is assumed to be rational. It is the rational agent, who often is used as the representative agent of an economy. Here we have a mouthful which requires a bit more background. Let us explain.

Economics in the twentieth century was driven to a large extent by contributions from Nobelist John von Neumann and Oskar Morgenstern [16], and Nobelists Gérard Debreu [17], Maurice Allais [18], and Kenneth Arrow. A beautiful collection of key authors who were instrumental in establishing economics as a science is contained in Chichilnisky [19]. In the development of economic theory, many intellectual resources were allocated to erecting a formal edifice of economic decision-making. Two important underlying conditions in formalizing will be the existence of rationality and the assumption we can model an economy via the idea of representative agent. The mathematically driven interpretation of economics avoids getting into formalizing intricate relations between individuals. Hence, the use of the representative agent concept. The representative agent is deemed to be rational. A fundamental driving assumption, proposed by amongst others, Nobelist Thomas Sargent [20], is that of 'rational expectations'. In essence, a representative agent does exhibit rational expectations if they adjust expectations for the future (say, the price of an asset) given the mistakes they made in past estimations. In other words, we learn from past mistakes.

We can say that decision makers amongst themselves will, after many adjustments on the basis of past information, converge toward having identical probability distributions on uncertain future prices. Naively, we can think of each decision maker forming his or her own subjective probability distribution over future asset prices. It is important to provide, at this point in our discourse, for a heuristic understanding of how we can view objective versus subjective probability. A good way to understand the fundamental difference between those two types of probability is provided for by the distinction between risk and uncertainty. Uncertainty, if assessable by some number, will be informed with the aid of subjective probabilities. As an example, can we objectively assess how many roofs will be blown off houses (in my neighbourhood) in the next ten years? We cannot. However, risk, if assessable by some number, will be informed with the aid of objective probabilities. In actuarial science, we can infer the risk of a certain type of car to be involved in an accident in a certain location. In snowy countries, insurance companies can assess the risk of an accident with an SUV as opposed to a regular sedan. We have frequencies of accident occurrences and from these, one can build objective probability distributions. The reader can sense that the distinction between risk and uncertainty also hints back to different types of information. In essence, uncertainty is not calculable, whilst risk is. It is not calculable because the information we have available is so random that it is unreliable.

Within rational expectations, when time goes by, and information is updated by economic actors, the subjective probability distributions start converting themselves into identical objective probability distributions. They become ultimately, objective, as all decision makers now share the same distribution. This idealization does not preclude random information. The assumption **does not say** that price behaviour is predictable. The unpredictability of future asset prices is intimately linked to nonrecurring information, that is, information which is not predictable.

We started out this discussion with the assertion that economics is preoccupied with giving insights into economic decision making. The rational representative agent used in macroeconomics, prevents researchers from having to deal with types of agents. Such behaviour (when types of agents were to be assumed) is typically more difficult to formalize. Let us give a simple example which relates back to our discussion on the maximization of utility. Let us think of nonrational behaviour. For instance, in behavioural economics 'noise traders' are an example of such nonrational agents: they trade on the basis of noise, that is, information which is not deemed fundamental. A noise trader may not follow the maximization assumption we mentioned earlier. Hence, if a market were to be characterized only by noise traders, the demand functions would reflect a decision-making process which is very difficult to model. Here is another example. Later on in the book, we will discuss some of the axiomatic frameworks of economic decision-making.

One intuitive condition is the so-called transitivity of preferences. If, as a consumer, I prefer good A over good B; and moreover, I prefer good B over good C, then by transitivity of preferences, I should prefer good A over good C. The quantities here are economically (and rationally) determined. Deviation from transitivity leads to deviant decision-making which is very hard to formalize. Those impairments toward proper formalization can be very deep. A more advanced example can refer to some of the very well documented paradoxes in axiomatic decision-making such as the Allais [21] and Ellsberg [22] paradoxes. We shall come back later to those paradoxes.

1.3.3 More on Markets: the Top-Down versus Bottom-Up Approaches

In recent years, economics has also opened doors to what one could call the 'individual' agent approach. In very simple terms, we could distinguish two grand groups of approaches in economic theory. We like to call this the opposition between 'top-down' versus 'bottom-up' modelling. In the top-down approach, the intricate relations between economic agents is specifically not modelled. Rational expectations and the representative agent approach are examples. In the bottom-up approach, it is precisely the opposite: one starts from the agent and a model is created on the basis of the 'type' of interaction of individuals. Behavioural economics/finance are examples of such an approach. Here is an example. When the UK government 'nudged' economic actors into complying to pay taxes, it meant essentially that the UK government was trying to induce a degree of compliance amongst taxpayers via psychological means: that is, "your neighbour is paying all his/her taxes - why do you not?" Nudging, as it is called, is very difficult to describe in formal terms. But it could be regarded as an essential part of running an economy. Behavioural economics does allow to look for behaviour which can characterize groups of economic actors. The many effects that have been tallied in behavioural economics when observing stock market interventions are another example. The so-called 'Monday' effect which attempts to document negative stock returns on a Monday, shows behaviour which is difficult to explain from an axiomatic, representative agent, decision-making point of view.

1.3.4 The Role of Probability in Economics

From the preceding, we can sense that although economics, on prima facie, can be looked upon as quite model-driven, it almost has to be, by default. Economics as a science would struggle to augment our understanding of human decision-making if it were solely descriptive. In that regard, economics does echo physics. However, we need to be careful. It is important to keep in mind though, how we treat probabilities in economics as opposed to how we treat them in physics.

The concept of probability is a central concept in economics, certainly within the realm of microeconomics. Economics traditionally makes a distinction between subjective; objective and so-called mixture probabilities (which contain both objective and subjective probabilities). The central decision-making axiomatics in economics work along those three divisions. Probability, within statistical mechanics and quantum theory plays a very important role too. The deep philosophical underpinnings which not only explain the need for probability but also the types of probability, in both disciplines, provide for an important debate which will inform even better the important variable-mapping exercise which we shall undertake in this book.

In economics, quantities, such as the price as the outcome of the market, form part of probabilistically measured events. Those events have been rendered by a human construction: namely, the market. The economic quantities, like price, mimic the physical quantities we can observe in nature, often via human construction too. Whilst in economics the human construction is the market; in physics the human construction which aids in determining a physical quantity is often a machine: the Geiger counter or the oscilloscope or the accelerometer, to name a few. In physics, an experiment studies the behaviour of particles under the influence of potentials, fields, and so on. A market can be understood as a setup where agents interact with their surroundings and this interaction can influence their decisions. We have provided for examples earlier, how we can formalize this interaction, when, for instance, we adhere to the rational expectations assumption. A price is an economically determined quantity which has coordinates living in some space. The price can have the coordinates of $x and a time variable, for instance. We can think of the price in other ways too – for example, it can represent the decision maker's 'position' in an appropriately defined 'economic' space. More on this later.

The structure of causality we mentioned when discussing the general form of physics frameworks, is much more delicate in economic systems. Given the widespread absence of laws in economics, pinpointing a structure of causality is fraught with difficulties. It is precisely this argument which can be invoked to motivate our exploration of bringing physics-inspired approaches to economics. It might help provide a causal framework which is consistent with the two main missions of economics we mentioned earlier, that is, understanding economic decision-making and markets.

1.3.5 More on Markets: Some Additional Thoughts

Markets can only exist when there is a symbiotic and (often) very complex relationship between agents and the decision-making those agents perform. The decision-making is often information-based and the types of information which drive the decision-making will also be an object of discussion later in the book. We hasten to add, that markets can, to a degree, function without much governmental intervention. However, sometimes such intervention is of key importance. The setting of interest rates in the economy, as an example, is directed by monetary policy. In the absence of such policy, interest rates would be very random and could have hugely damaging effects on the economy. It is true that the underlying 'glue' of much market functioning is provided by a well-known principle known as the 'invisible hand' (see Ullman-Margalit [23]) that Adam Smith devised in the eighteenth century. It is a cornerstone idea of economics. But it is by no means the only driver of a successful market.

Whilst interest rates benefit from governmental intervention, stock markets may well provide for a counterexample. Current stock markets are driven to a very large extent by types of information whether public; semiprivate or private information and their dynamics occurs in the absence of much governmental intervention. Granted, there are basic regulators. The SEC, the US Securities and Exchange Commission, steers stock markets and minimizes corrupt behaviour as much as possible.

Even though past price history may be a predictor of the future, this is often not at all the stance academia will subscribe to. As an example, the price foundation formalism in option pricing follows the physics-based Brownian motion and the random driver process in such a motion is known as the Wiener process, named after polymath Norbert Wiener. In physics analogy, the dynamical evolution of the particle is now emulating the dynamical evolution of the price. It is clearly the human mind, via the construction of a market, which renders the 'price' as an object that lives in an economic space analogous to the real space for physical objects. Given the origins of Brownian motion, that is, of Newtonian physics, it is clear that such physics can underpin the dynamical structure of price behaviour. This is not an unfounded statement. Option pricing theory uses precisely Brownian motion as the model to emulate dynamic asset price behaviour. We can safely claim that option pricing theory is amongst the most successful theories ever devised in economics and finance and it lies at the root of a multi-trillion-dollar market. The option pricing market has come into being because of the construction of a theory: option pricing theory. The financial reality in some sense, was created out of a theory. We invite the reader to ponder over this statement, see [24] for an excellent account.

Can we say that a somewhat similar situation may occur in case of a physics-like theory of finance? Is a question like "Does a physical theory give meaning to the motion of electrons in a cloud chamber in terms of the *notion of trajectories*?" relevant and useful here? Following Einstein – can we say that it is the theory which decides what we can observe?

1.4 About the Book

We began with a discourse on some of the milestones from the history of the developments in physics (Section 1.1). These milestones indicated how the evolution of physics from its ancient form of natural philosophy to a more general, modern form of natural philosophy *and* natural science took place. We then proposed that the framework of any physical theory of motion could be considered as built out from an inseparable combination of a specific dynamical paradigm and an assumed geometric structure of space-time, and it provided a plausible explanation of our proposal by analyzing a simple dynamical situation (Section 1.2). Hopefully, these discussions also shed light on the fact that the subject matter of physics, in general, has not just mathematical but also philosophical, interpretational, and experimental findings-based ingredients. In the previous Section (1.3) we highlighted that the economics-financial processes originate from the behaviour (in terms of choices, preferences, or decisions) of its agent-systems. Before moving on, we here summarize the contents of the next chapters.

Our discussion from Section 1.2 continues in Chapter 2, which serves more as a compendium of selected elements of classical and quantum mechanics. Specifically, it provides an outline of the analytical formulation of classical mechanics, explains some of the important aspects of the notion of an 'ensemble' and gives an overview of some selected topics from the formulation of quantum theory, such as the wave function, operators, uncertainty principle, expectation values, and wave packets. This chapter may help the readers from the social sciences to gain certain familiarity with the basic formulation of the classical and quantum dynamical frameworks before getting into the quantum-like (Part II) and the quantum physics-like (Part III) modelling approaches.

Continuing from Section 1.3, Chapter 3 aims to provide for a brief appreciation of some of the key events in the development of economic science. We emphasize the idea of the 'observable' in that chapter and the quest for making economic theories to be falsifiable. It is often misunderstood that economics is merely a theoretical construct, which is in essence not even deemed falsifiable. Chapter 3 tries to dispel this myth, albeit in a somewhat shallow way. We briefly discuss the important impact econometrics is playing in economics, and also emphasize the lineup of parsimoniously selected variables being inputs in one (if not the 'one') most successful theory in asset pricing, that is, the Black–Scholes option pricing theory.[15] In essence, this theory does not appeal to any concepts which almost by definition are unobservable, like the idea of utility. In effect, the theory is generated in a so-called risk-neutral world. This is not to say the unobservable concepts like utility are of no use. The mathematical economics edifice built up in the twentieth century laid the

[15] We note that in the basic setup, the Black–Scholes model assumes the volatility parameter to be constant.

groundwork for many extremely important, formal, decision-making frameworks all relying on expected utility expressed via different probabilistic frameworks. We cover those frameworks in some detail in Section 5.4.

Chapter 3 also delves somewhat into the 'novel arrivals' of economics, so to speak. Behavioural and experimental economics is discussed and we try to highlight that especially in those two new fields, the experimental verification of some of the formal decision-making models has made phenomenal advances. As we will also discuss in more detail in Section 5.4, the divergence between observed decision-making relative to the purely theoretical constructs has called in a very fruitful development of alternative formal models. In the end, it has also opened the door to one of the key ideas in this book, namely, the use of quantum probability in decision-making.

Chapter 3 also considers some elementary ideas in computational and neuroeconomics and with the latter novel field, we do note the recent ideas on the 'quantum-ness' of the brain. We round off Chapter 3 with what we believe is a notion of fundamental importance, especially in fields like finance and economics: the value of information. If trillions of dollars are being traded on a daily basis, it is safe to say that modelling information and formalizing the impact such modelled information may have on dynamic price behavior is of paramount importance (but please see what follows for a discussion on so-called price formation). In a forthcoming Section 6.2, we will show how a particular measure of information can 'naturally' be embedded in a pricing framework which functions along quantum mechanical lines.

Chapter 4 delves into some of the highlights of financial research, at least in its connection with asset pricing. We make the distinction between so-called price discovery, which is highly dependent on information, and price formation which is rather linked to trading per se – and quite independent from information. We discuss, in some depth, the Kyle model in (Section 4.2) and the highly important concept of market depth. The Kyle model considers how fast private information enters price, whilst the Milgrom approach in Section 4.3 gives very interesting insights into the formation of bid and ask prices.

Chapter 4 then continues to bring forward some elementary ideas on arbitrage taking (using the least action idea from physics) and continuous time processes as they relate to stock price dynamics, volatility, and forward rates. Chapter 4 then proceeds with discussing the relation between liquidity and the type of information. We round off the chapter with some ideas on using quantum concepts relative to the Rosenow model. After discussing the Langevin equation which forms an integral part of that model, we ponder whether it may be palatable to argue for the existence of a 'force' which relates to a specific potential which we will be discussing in more depth in Chapter 6.

Chapter 5 has as its main objective to 'set the scene' so to speak, on why ongoing research, in applying formalisms to social science has provided for results which are noteworthy and important. The chapter starts out with a very practical objective: why quantum in social science? The quantum-like applications, as they are often called, come to the research scene (especially in decision-making), via one potent argument: the existence of incompatible observables in social science contexts. This also brings in a different characterization of uncertainty. As we mention in the beginning of Chapter 5, order effects when questioning experiment participants (and in real life) are almost ubiquitous. Compatible events require two conditions: (i) events need to be realized simultaneously and (ii) the order of events is irrelevant. It is well known that the order of events does play a role in repeated experiments in psychology. Order effects, that is, asking question 'A' first followed by question 'B', may give different results relative to the case where question 'B' is asked first and then followed by question 'A'. Moreover, the simultaneous realization of two events may well pose problems too in real life. As an example, if one acts joyously in one surrounding but then acts sad in a different surrounding, it is not possible to *simultaneously* observe both events. The idea of interference could arise from incompatibility and it has been shown that such interference can cause the so-called conjunction fallacy, which we discuss in Section 5.2.

Chapter 5 also surveys briefly some other contributions related to the disjunction fallacy and dynamic inconsistency. We then survey in a little more depth some of the mainstream expected utility frameworks, notably the Savage model (from where the conjunction fallacy emerges from). We also hint toward some of the alternatives which occurred in economic theory. Chapter 5 then completes with an appreciation of two milestone results in quantum-like research, the so-called QQ equality which is treated in Section 5.5 and a quantum-like extension of the Aumann theorem in Section 5.6).

In Chapter 6, we start out with attempting to define the 'reason of existence' of a wave function in social science. This is not a wave function, like in classical mechanics, but rather a quantum mechanical wave function. Chapter 6 then continues with introducing in Section 6.2 the Hawkins and Frieden approach which shows how Fisher information enters a Lagrangian. Quantization has an information theoretic basis. The chapter then continues to discuss how the quantum potential, a key concept in Bohmian mechanics, leads to defining so called Q-risk in a finance context. It also discusses how the forces derived from real and quantum potentials, when using financial return data, do behave differently within the walls of both potentials. We round off that chapter with a brief discussion on the path characterization (from a financial perspective) and the possible ideas which may emerge when one considers Bohmian mechanical ideas in the setting of a macroscopic fluid bath.

Chapter 7 starts out with a discussion of applications of a quantum harmonic oscillator which allows for a quantized version of supply and demand. We then move on to look at quantum probability within finance, using some cautionary note that there is (only) mere indirect evidence for the use of such a type of probability in finance. However, the treatment of uncertainty using the quantum framework is appealing in finance. As we remarked in Chapter 5, within decision making per se, the rationale for the use of quantum probability is quite firmly established.

The two next sections in that chapter focus on first applying quantum probability in an experimental setting where participants need to make investment decisions in a setting which is very much akin to that of the two-stage sequential gambling problem we discuss in Chapter 5. In the next section we give a brief overview of how quantum probability can be used in a setting where there is missing financial data.

Chapter 7 also looks at the use of the formalism of quantum field theory, that is, where thus wave functions are now fields and those fields are operators (which link to so-called creation and annihilation operators). We also provide for a discussion on open systems in social science (i.e. where interactions of a system with its environment (a bath) are now allowed. Non-Hermiticity in finance is then discussed and a key result is the observation that the Hamiltonian derived from Black–Scholes option theory is not Hermitian. The key argument revolves around the absence of arbitrage which is a condition of paramount importance in any asset pricing model (such as options).

We round off the chapter with two discussions. One discussion on a general platform of models which are anchored around two parameters: (absence) existence of memory and (a) symmetry of time. A central approach in this discussion revolves around the so-called stochastic approach to quantum mechanics. The very last section of that chapter then argues for some ideas around whether there are parallels between the way that formalisms are created in quantum physics and quantum-like formalisms are constructed in social science.

From Chapter 8 onwards (i.e. in Part III) we set out to explore the analogy between –

- the *behaviour of agents* belonging to the *economics/finance world* and
- the *behaviour of systems* belonging to the *physical world.*

Specifically, we investigate the extent to which such an analogy could be precisely developed and the way in which the resulting quantum physics–analogous models may prove beneficial to the economic sciences. In similarity with the physical theories of motion, our approach leads to a theory of 'motion' for the agents of an economy and also, to help view the economics-finance world as consisting of causally functioning systems. With the aim of arriving at some definite guidelines on constructing *quantum* physics–analogous models of economics and finance, in Chapter 8 we begin with a discussion on what we prefer to call the *physical* foundations of quantum theory and then proceed to derive expectations from the resulting models that would be constructed by drawing precise analogies from the models of quantum physics. To make the 'analogy' exploration more systematic, we also briefly review a variety of criteria put forward by the philosophers of science to evaluate analogical arguments.

In Chapter 9, we choose 'electron two-slit interference' as a specific quantum physical situation and provide some (nondynamical) preliminaries of the corresponding model in terms of Dirac's bra-ket symbols. Next, we propose an 'analogy checklist' of notions explaining a set of physics–like kinematic and dynamic quantities that would be required as ingredients for constructing quantum physics–like models for economics and finance. We then consider a situation involving the seemingly stochastic transitions of an asset price and provide a sketch of its model bearing analogy with the electron two-slit interference situation. We call it the 'agent two-preference interference' model.

We continue in Chapter 10 with the quantum dynamical aspects of the electron two-slit interference situation and by precisely following it, develop further, in an analogous manner, the dynamical agent two-preference interference model. However, unlike the usual discussions of the electron two-slit interference situation, which consider the motion of the electrons from the slits to the screen as 'free' from any external influences (potentials), we do not restrict the analysis to the $V = 0$ case. Rather, we analyze the electron two-slit interference case for a nonzero V and analogously investigate further the agent two-preference interference situation.

Chapter 11 provides simulations of Bohmian trajectories and probability distributions for a variety of parameter combinations. It ends by providing physics-analogous interpretations of the potential energy functions and also, of the other model parameters.

Finally, we remark that, no exploration is ever complete and there is always something left out for future work. In Chapter 12, we give an outline of some future directions that could possibly emerge from this novel ('physics-analogous') approach to modelling, discussed in Part III.

References

[1] Walker, J. (2007). *The Flying Circus of Physics*. New York, NY: John Wiley.

[2] Feynman, R. P., Leighton, R. B. and Sands, M. (1965). *The Feynman Lectures on Physics, volume 1*. New York, NY: Basic Books.

[3] Singer, C. J. (1941). *A Short History of Science to the Nineteenth Century*. Oxford: Clarendon Press.

[4] Modinos, A. (2014). *From Aristotle to Schrödinger*. Cham: Springer.

[5] Penrose, R. (2004). *The Road to Reality*. London: Jonathan Cape.
[6] Baggott, J. (2011). *The Quantum Story*. Oxford: Landmark Science.
[7] Einstein, A., Podolsky, B. and Rosen, N. (1935). Can quantum-mechanical description of physical reality be considered complete? *Physical Review*, 47, 777–780.
[8] Bohm, D. (1951). *Quantum Theory*. Englewood Cliffs, NJ: Prentice-Hall.
[9] Clauser, J. F., Horne, M.A., Shimony, A. and Holt, R.A. (1969). Proposed experiment to test local hidden-variable theories. *Physical Review Letters*, 23, 880–884.
[10] Athalye, V. (2025). Unifying Perspectives: All-Encompassing Reality vs. All-Inclusive Physics. *Journal of Physics: Conference Series*, 2948: 012012.
[11] Bialek, W. (2012). *Biophysics: Searching for Principles*. Princeton, NJ: Princeton University Press.
[12] Tuszynski, J. A. (Ed.) (2006). *The Emerging Physics of Consciousness*. Berlin: Springer.
[13] Anderson, E. (2017). *The Problem of Time*. Cham: Springer.
[14] Greiner, W. (2004). *Classical Mechanics: Point Particles and Relativity*. New York: Springer International.
[15] Varian, H. (1992). *Microeconomic Analysis*. 3rd Ed., New York, NY: W. W. Norton.
[16] von Neumann, J. and Morgenstern, O. (1944). *Theory of Games and Economic Behavior*. Princeton, NJ: Princeton University Press.
[17] Debreu, G. (1959). *Theory of Value: An Axiomatic Analysis of Economic Equilibrium*. New Haven, CT: Yale University Press.
[18] Allais, M. and Hagen, O. (Eds.) (1979). *Expected Utility Hypotheses and the Allais Paradox: Contemporary Discussions of Decisions under Uncertainty with Allais' Rejoinder*. Theory and Decision Library. Vol. 21. Dordrecht: Springer.
[19] Chichnilsky, G. (Ed.)(1999). *Markets, Information and Uncertainty: Essays in Honor of Kenneth. J. Arrow*. Cambridge: Cambridge University Press.
[20] Sargent, T. J. (2013). *Rational Expectations and Inflation*. Princeton, NJ: Princeton University Press.
[21] Allais, M. (1953). Le comportement de l'homme rationnel devant le risque. *Econometrica* 21(4), 503–546.
[22] Ellsberg, D. (1961). Risk, ambiguity and Savage axioms. *Quarterly Journal of Economics*, 75, 643–669.
[23] Ullmann-Margalit, E. (1978). Invisible-hand explanations. *Synthese*, 39(2), 263–291.
[24] MacKenzie, D. and Millo, Y. (2005). Negotiating a Market, Performing a Theory: The Historical Sociology of a Financial Derivatives Exchange. *American Journal of Sociology*, 109, 107–145.

2

Elements of Classical and Quantum Mechanics

In this chapter we attempt to expound on basic and essential ideas (for further use in the book) from both classical and quantum mechanics. The notion of 'causality' and the idea of 'ensemble' will be of central importance in the book. The chapter is somewhat technical in nature but only requires an elementary knowledge of calculus.

The first three sections take a review of some of the elements of classical mechanics, whereas the remaining sections are devoted to the elements of quantum mechanics. We assume the reader's familiarity with at least the basic (if not advanced) formulations of these theories. Therefore, our handling of the corresponding topics will be far from thorough. The purpose of the material presented here is to serve only as a quick reference and background for the discussions appearing in Parts II and III. For detailed treatments of these theories, we refer the reader to excellent accounts such as those of Goldstein [1], Holland [2], and Bransden and Joachain [3]. Specifically, the material presented here can be seen as pointing out the central roles played by the notion of 'causality' and the idea of an 'ensemble' in the mechanistic settings owing to their anticipated relevance in the construction of quantum-like and quantum physics–like models for the economic sciences. This way, the present chapter could also be seen as an elaboration of the background given by Athalye and Haven in [4].

In classical mechanics, the instantaneous configurations of a physical system having n degrees of freedom are described in terms of points $q \equiv (q_1, \cdots, q_n)$ belonging to an n-dimensional (n-D) Cartesian hyperspace, which is also called the configuration space. The coordinates $q_1, \cdots, q_n$ are called the n *generalized* coordinates. The trajectory of the system is nothing but a curve traced by the system point q by virtue of the motion of the system in the configuration space between the time instants t_1 and t_2.

In general, the configuration space does not refer to the physical three-dimensional space, and the generalized coordinates are not necessarily the position coordinates. However, in the following, wherever the dynamical situations involving the motion of a system (modelled as a particle) of mass m taking place in the real physical space are referred to, we prefer to use the $\mathbf{x}$-notation (or simply x for one-dimensional problems) for the system point, over the (usual) q-notation and consider its Cartesian (i.e. $\mathbf{x} \equiv (x, y, z)$) representation, with x, y, and z as the position coordinates.

2.1 The Euler–Lagrange and the Hamilton–Jacobi Equations

In Newtonian mechanics the cause behind the *change* in the state of motion (i.e. the 'dynamics') of the system in question is represented by an external 'force' acting on the system. As mentioned in Section 1.2, such an external force, according to Newton's second law of motion, is given by Equation (2.1):

$$\mathbf{F} = \frac{d\mathbf{p}}{dt}. \tag{2.1}$$

In other words, Newton's second law identifies the external force $\mathbf{F}$ as the *cause* behind the change in the state of motion of the system in question. If $\mathbf{F}$ is a conservative (pure position-dependent) force[1], it can be expressed as the negative gradient of some potential energy V:

$$-\nabla V = \mathbf{F}. \tag{2.2}$$

Let us turn to the analytical formulation of Newtonian mechanics.

For a system of mass m moving with reference to a $(3+1)$-dimensional Galilean space-time – or, more generally, a Galilean–Newtonian space-time (see [5]) – along a trajectory $\mathbf{x} = \mathbf{x}(t)$ with the velocity $\dot{\mathbf{x}} = d\mathbf{x}/dt$, under the influence of its environment, which can be modelled in terms of the classical potential energy V, the Lagrangian is given by the *difference* between its kinetic and potential energies:

$$L(\mathbf{x}, \dot{\mathbf{x}}, t) = \frac{1}{2}m\dot{\mathbf{x}}^2 - V(\mathbf{x}, t). \tag{2.3}$$

In terms of its Lagrangian, the momentum $\mathbf{p}$ $(= m\dot{\mathbf{x}})$ of the system is given by

$$\mathbf{p} = \frac{\partial L}{\partial \dot{\mathbf{x}}}. \tag{2.4}$$

The Hamiltonian of the system is defined as a function on the six-dimensional phase space (x, y, z, p_x, p_y, p_z) by a Legendre transformation:

$$H(\mathbf{x}, \mathbf{p}, t) = \mathbf{p} \cdot \dot{\mathbf{x}} - L. \tag{2.5}$$

[1] A conservative force is the one for which the total work done in moving a system between two points is path-independent.

Using (2.3) and (2.5), the Hamiltonian can be seen as the *addition* of the kinetic and the potential energies of the system (with $|\mathbf{p}| = p$):

$$H = \frac{p^2}{2m} + V(x, t). \tag{2.6}$$

In analytical mechanics, the cause behind the *change* in the state of motion of the system is expressed in terms of the Hamiltonian of the system. It can thus be said that the Hamiltonian dictates the time-evolution of the state of the system (see the following discussion).

For simplicity, consider the motion of the system in the $(1 + 1)$-dimensional space-time. Also, consider two points: (x_0, t_0) and (x, t) and the *paths* that may join them. Then, according to Hamilton's principle, the actual path traversed by the system is that for which the action integral:

$$I(x, t; x_0, t_0) = \int_{x_0,t_0}^{x,t} L(x, \dot{x}, t)\, dt \tag{2.7}$$

has a stationary value. The value for the action integral being stationary means, in simple terms, that we would have obtained the same value even if an infinitesimally differing path joining the same end points were considered. Thus, the motion of the system between two fixed space-time points will be such that the variation in the action integral – due to the 'first-order' variation in the path – is zero:

$$\delta I(x, t; x_0, t_0) = \delta \int_{x_0,t_0}^{x,t} L(x, \dot{x}, t)\, dt = 0. \tag{2.8}$$

Hamilton's principle [(2.7) and (2.8)] implies the Euler–Lagrange equation of motion:

$$\frac{d}{dt}\frac{\partial L}{\partial \dot{x}} - \frac{\partial L}{\partial x} = 0. \tag{2.9}$$

Substituting L from (2.5) in (2.7), we can write the action function as

$$I(x, t; x_0, t_0) = \int_{x_0,t_0}^{x,t} (p\dot{x} - H)\, dt. \tag{2.10}$$

Hamilton's equations of motion can be obtained by varying (2.10) w.r.t. the position and momentum variables:

$$\dot{x} = \left(\frac{\partial H}{\partial p}\right)_{x=x(t),\, p=p(t)}, \tag{2.11}$$

$$\dot{p} = -\left(\frac{\partial H}{\partial x}\right)_{x=x(t),\, p=p(t)}. \tag{2.12}$$

Suppose the coordinates x, p are transformed to another pair of coordinates X, P, such that the new Hamiltonian is $K = K(X, P, t)$. Such a phase-space coordinate transformation is called *canonical* if Hamilton's equations (2.11)–(2.12) retain their form under the transformation, that is,

$$\dot{X} = \left(\frac{\partial K}{\partial P}\right)_{X=X(t),\, P=P(t)}, \tag{2.13}$$

$$\dot{P} = -\left(\frac{\partial K}{\partial X}\right)_{X=X(t),\, P=P(t)}. \tag{2.14}$$

Equations (2.13) and (2.14) can be obtained by a variational principle applied to an action function of the form (2.10):

$$I = \int (P\dot{X} - K)\, dt. \tag{2.15}$$

The integrands in (2.10) and (2.15) will then differ by a total time derivative:

$$P\dot{X} - K = p\dot{x} - H - \frac{dF}{dt}. \tag{2.16}$$

The function F is an arbitrary differentiable function of time, and of old and new coordinates. In particular, for $F \equiv F(x, X, t)$, from (2.16) we deduce that

$$p = \partial F/\partial x, \tag{2.17}$$

$$P = -\partial F/\partial X, \tag{2.18}$$

and

$$K = H + \partial F/\partial t. \tag{2.19}$$

If the functional form of F is known, the canonical transformation '$X = X(x, p, t)$; $P = P(x, p, t)$' can be constructed. In this sense, F is called the generating function of the canonical transformation.

The system Hamiltonian is the generator of infinitesimal canonical transformation, which is nothing but the physical evolution of the generalized canonical coordinates of the system during dt, an infinitesimal time interval. An arbitrary canonical transformation can be built from a succession of infinitesimal transformations. Therefore, the actual motion of the system over a finite time interval governed by Hamilton's equations is itself a continuous canonical transformation:

$$x = x(x_0, p_0, t), \qquad p = p(x_0, p_0, t). \tag{2.20}$$

Thus, a particular way of solving the problem of motion is to obtain the canonical transformation $(x, p, t) \longrightarrow (x_0, p_0, t_0)$, with x_0 and p_0 as constant coordinates and

t_0 is initial time. From Equations (2.13) and (2.14), it can be seen that this will be achieved if $K = 0$, giving

$$\dot{X} = \dot{x_0} = 0, \qquad \dot{P} = \dot{p_0} = 0. \tag{2.21}$$

From Equation (2.19) we get that for $K = 0$, F must satisfy the equation $\partial F/\partial t + H = 0$ and for this case, it is denoted by S. This way, we obtain what is known as the Hamilton–Jacobi equation:

$$\frac{\partial S}{\partial t} + H = 0, \tag{2.22}$$

where the Hamiltonian is given by

$$H = (\nabla S)^2/2m + V, \tag{2.23}$$

with the momentum $\mathbf{p} = \nabla S$, which is simply $\frac{\partial S}{\partial x}\hat{i}$ in one dimension (1-D) [see Equation (2.17)].

2.2 The Idea of an Ensemble in the Classical Context

A complete integral of the Hamilton–Jacobi equation is required for solving a dynamical problem in classical mechanics. In general, if the system under consideration has n degrees of freedom, then the complete solution $S \equiv S(\mathbf{x}, \boldsymbol{\alpha}, t)$ will involve n nonadditive constants (as the components of the vector $\boldsymbol{\alpha}$) of such an integral (see [6]). Even if the initial positions and momenta are fixed, a given Hamilton–Jacobi equation can admit different 'complete integrals', implying that the S-function corresponding to a given mechanical problem (i.e. with a prescribed Hamiltonian, $\mathbf{x}_0$ and $\mathbf{p}_0$) is not unique. The nonuniqueness of S actually allows infinitely many potential trajectories, with each one pursued by an element (i.e. an individual system) from an *ensemble* of identical systems [see chapter 2 of [2] for details].

To explain this, we consider an example of the motion of a free particle ($V = 0$), for which the Hamilton–Jacobi equation becomes

$$\partial S/\partial t = -(\nabla S)^2/2m, \tag{2.24}$$

which integrates to give

$$S(x, y, z, p_x, p_y, p_z, t) = -(1/2m)(p_x^2 + p_y^2 + p_z^2)t + xp_x + yp_y + zp_z, \tag{2.25}$$

where the components of the momentum $\mathbf{p} \equiv (p_x, p_y, p_z)$ are the nonadditive constants of the complete integral of (2.24). The particle trajectory is given by

$$\mathbf{x} = \mathbf{x}_0 + \mathbf{v}t. \tag{2.26}$$

Keeping the momentum $\mathbf{p}$ fixed and varying $\mathbf{x}_0$, Equation (2.26) can be understood as describing an ensemble pursuing a set of parallel uniform rectilinear motions. The free particle action function can be obtained in one more way, by using the free-particle Lagrangian

$$L = \frac{1}{2}m\dot{\mathbf{x}}^2 \tag{2.27}$$

and integrating along the trajectory (2.26):

$$S(\mathbf{x}, t; \mathbf{x}_0, 0) = (m/2t)(\mathbf{x} - \mathbf{x}_0)^2. \tag{2.28}$$

Again, the function (2.28) describes an ensemble of particles all originating from the point $\mathbf{x}_0$ with a range of momenta $\mathbf{p}$. The motions generated by (2.25) and (2.28) coincide when $\mathbf{x}_0$ and $\mathbf{p}$ are chosen to be the same in both cases. For S of the form (2.25) we have

$$m\dot{\mathbf{x}} = \nabla S = \mathbf{p}. \tag{2.29}$$

With the initial condition $\mathbf{x}(0) = \mathbf{x}_0$, this integrates to give (2.26). For S of the form (2.25) we have

$$m\dot{\mathbf{x}} = \nabla S = m(\mathbf{x} - \mathbf{x}_0)/t. \tag{2.30}$$

This also integrates to give (2.26) if $\mathbf{x}(0) = \mathbf{x}_0$. The different functional dependence of ∇S can be easily noted: In Equation (2.29), it is constant in space and time, whereas in Equation (2.30), it is a variable.

2.3 Continuity Equation for Particle Density

The theory of classical mechanics implies a *deterministic and predictable* system motion. This is in the sense that, using the laws of classical mechanics, a precisely specified initial state $[\mathbf{x}(0), \mathbf{p}(0)]$ of the system can be precisely mapped into a final state $[\mathbf{x}(t), \mathbf{p}(t)]$. However, in practice, our knowledge of the initial position and momentum of the system is limited. Such a limitation has led to the idea of modelling the system of interest in terms of a set of *identical, noninteracting* systems – called the Gibbsian-type ensemble – having one-to-one correspondence with a range of initial conditions and hence, with a set of system motions, such that each individual motion takes place in accordance with the underlying Newtonian causal dynamics. This is the basis on which the theory of classical *statistical* mechanics has been built, which implies a *deterministic, but not predictable* system motion. By virtue of being deterministic, both classical mechanics and classical statistical mechanics respect the principle of causality.

As seen in the previous section, the Hamilton–Jacobi formulation of classical mechanics naturally admits an ensemble description. Solutions of the following equation

$$\mathbf{v} = \dot{\mathbf{x}}, \tag{2.31}$$

with

$$\mathbf{v}(\mathbf{x}, t) = \frac{\nabla S(\mathbf{x}, t)}{m} \tag{2.32}$$

represent an ensemble of possible system motions $\mathbf{x}(t)$. The motion of each system from the ensemble is completely determined by the (common-to-all) S function and the initial position of the system, see [2] for details.

For such an ensemble of systems, in classical statistical mechanics, a density function, $\rho = \rho(\mathbf{x}, t)$, is introduced, which gives the number of individual systems per unit volume. According to the Gibbs approach, such a function represents a variety of potential individual system motions. Its time-evolution, subject to the local conservation of the ensemble elements (systems), is given by the continuity equation:

$$\frac{\partial \rho}{\partial t} + \nabla \cdot (\rho \mathbf{v}) = 0. \tag{2.33}$$

We can interpret the function $\rho(\mathbf{x}, t)$ as representing 'probability density' (i.e. probability per unit volume) as well, so that the probability that an individual system lies between the points $\mathbf{x}$ and $\mathbf{x} + d\mathbf{x}$ at time t is given by $\rho\, d^3x$. Moreover, since the system must exist somewhere in space, it is sensible to require that

$$\int_{\text{all space}} \rho\, d^3x = 1. \tag{2.34}$$

2.4 Classical–Quantum 'Mash-up'

Just as the Hamilton–Jacobi equation (HJE) is one of the central dynamical equations of the analytical formulation of classical mechanics, the time-dependent Schrödinger equation (SE)

$$i\hbar \frac{\partial \psi}{\partial t} = \left(-\frac{\hbar^2}{2m} \nabla^2 + \hat{V} \right) \psi = \hat{H} \psi \tag{2.35}$$

is the central dynamical equation of nonrelativistic quantum mechanics. Thus, SE is a statement that the operator

$$\hat{H} = \left(-\frac{\hbar^2}{2m} \nabla^2 + \hat{V} \right) \tag{2.36}$$

is operationally equivalent to the operator $i\hbar\frac{\partial}{\partial t}$ and is called the Hamiltonian (or the total energy) operator, and $\psi \equiv \psi(\mathbf{x}, t)$ is the (complex) wave function, which is also called the 'state' function of the system in question.[2]

Several authors have investigated the connection between the HJE and the SE, mainly to explore the origin of the latter (see e.g. [6], [7] and the references therein; also, see [2]). In the following we give some remarks that make the connection between the two formulations (classical and quantum mechanics) apparent. We begin by expressing ψ in the polar form:[3]

$$\psi = R\, e^{iS/\hbar} = R\cos(S/\hbar) + iR\sin(S/\hbar). \tag{2.37}$$

If we substitute it in Equation (2.35) and separate the real and imaginary parts, we obtain two equations – one following after equating the real parts:

$$\frac{\partial S}{\partial t} + \frac{(\nabla S)^2}{2m} + V - \frac{\hbar^2}{2m}\frac{\nabla^2 R}{R} = 0 \tag{2.38}$$

and the other after equating the imaginary parts:

$$\frac{\partial R^2}{\partial t} + \nabla \cdot \left(\frac{R^2 \nabla S}{m}\right) = 0. \tag{2.39}$$

Let us compare the preceding pair of equations with the Hamilton–Jacobi (2.22) and the continuity (2.33) equations:

$$\frac{\partial S}{\partial t} + \frac{(\nabla S)^2}{2m} + V = 0 \tag{2.40}$$

and

$$\frac{\partial \rho}{\partial t} + \nabla \cdot \left(\frac{\rho \nabla S}{m}\right) = 0. \tag{2.41}$$

It can be seen that Equation (2.39) is exactly similar to Equation (2.41), with the 'classical' probability density (ρ) replaced by the modulus-square of the wave function (or the square of the amplitude of the wave function), $|\psi|^2 = R^2$, to be later on identified as the 'quantum' probability density. Equation (2.39) is thus known as the continuity equation for the 'quantum' probability density. Equation (2.38) is called the 'quantum' Hamilton–Jacobi equation. It differs from Equation (2.40) with respect to the fourth term in its LHS. It is called the quantum potential energy (denoted by Q):

$$Q = -\frac{\hbar^2}{2m}\frac{\nabla^2 R}{R}. \tag{2.42}$$

[2] with $\left(-\frac{\hbar^2}{2m}\nabla^2\right)$ as the kinetic energy operator and $\hat{V}$ as the potential energy operator, see Section 2.6.

[3] This is done under the Wentzel-Kramers-Brillouin (WKB) approximation or the Liouville-Green (LG) method, a method for the approximate treatment of the Schrödinger wave equation. See e.g. [8].

The pair $(S, -\rho)$ of classical scalar functions turns out to be a pair of canonically conjugate variables. We can transform this pair to another pair, that of classical complex wave functions (ψ_{cl}, ψ_{cl}^*) [where ψ_{cl}^* is the complex conjugate of ψ_{cl}] and can express the function ψ_{cl} in the polar form:

$$\psi_{cl} = R_{cl}\, e^{iS_{cl}/A}, \tag{2.43}$$

with A as a constant having the dimensions of action and write its amplitude as

$$R_{cl} = |\psi_{cl}| = +\rho^{1/2}. \tag{2.44}$$

The classical wave function ψ_{cl} satisfies the following nonlinear wave equation:

$$iA\frac{\partial \psi_{cl}}{\partial t} = \left(-\frac{A^2}{2m}\nabla^2 + V + \frac{A^2}{2m}\frac{\nabla^2 R_{cl}}{R_{cl}}\right)\psi_{cl}, \tag{2.45}$$

with a similar equation for ψ_{cl}^* [see [2] for details]. The preceding equation looks very much like the time-dependent Schrödinger equation (2.35), except for the third term on the RHS. We may call Equation (2.45) the classical Schrödinger equation. Due to the presence of the third term in its LHS, it is nonlinear in ψ_{cl}, unlike the usual (quantum) Schrödinger equation, that is, Equation (2.35), which is linear in ψ. Thus,

- the pair 'classical Hamilton–Jacobi equation – classical continuity equation' leads – by means of a canonical transformation $(S, -\rho) \longrightarrow (\psi_{cl}, \psi_{cl}^*)$ – to the *classical* Schrödinger equation; and
- the *quantum* Schrödinger equation leads – by assuming its solution in polar form – to the pair 'quantum Hamilton-Jacobi equation – quantum continuity equation.'

It is interesting to note that the absence of the term '$-\frac{\hbar^2}{2m}\frac{\nabla^2 R}{R}$' in the quantum Schrödinger equation corresponds to its presence in the quantum Hamilton–Jacobi equation, whereas the absence of the term '$+\frac{A^2}{2m}\frac{\nabla^2 R}{R}$' in the classical Hamilton–Jacobi equation corresponds to its presence in the classical Schrödinger equation.

The quantum Hamilton–Jacobi equation leads to the quantum generalization of Newton's second law,

$$\mathbf{F} = m\ddot{\mathbf{x}} = -\boldsymbol{\nabla}(V + Q), \tag{2.46}$$

by means of which the idea of 'force' acting on a system and the notion of space-time 'trajectory' of a system could be restored and the working of the principle of 'causality' could be made transparent in the quantum paradigm. This is done in the ontological interpretation of quantum theory (see [9]).

All these remarks indicate that there is somehow an intricate connection between the classical and the quantum dynamical structures.

2.5 The Idea of an Ensemble in the Quantum Context

The idea of an ensemble has already been introduced in Section 2.2 in the classical context. Here we discuss the same concept in the quantum mechanical context. In quantum theory, the idea of an ensemble is more intricate than that emerging in the analytical formulation of classical mechanics and its statistical extension. A detailed discussion would ideally point out not just the similarities but also the essential contrasts between the ways the two formulations treat the idea of an ensemble. However, such a discussion would go beyond the scope of this section, whereas here we intend to provide just an introduction to the basic idea of an ensemble. For details, we refer the reader to [10]; see also [2].

For our purpose here, we follow the de Broglie–Bohm (dBB) formulation of quantum theory. According to the dBB formulation, an individual physical system is not just a particle but both *a particle and a wave*, with the latter guiding the motion of the former. The propagation of the wave in space-time is described in terms of its wave function $\psi(\mathbf{x}, t)$, which is a solution to the Schrödinger equation (2.35), whereas the trajectory, $\mathbf{x}(t)$, of the particle aspect of the system in space-time, is obtained as the solution to the following equation:

$$\dot{\mathbf{x}} = (1/m)\nabla S|_{\mathbf{x}=\mathbf{x}(t)}, \tag{2.47}$$

where S is the phase of the wave function when expressed in the polar form; see Equation (2.37). The pair of variables $[\psi(\mathbf{x}, t), \mathbf{x}(t)]$ describes the 'state' of an individual physical system capable of exhibiting both wave-like and particle-like behaviours. Starting from Equation (2.38), it can be seen that, just as in the classical case, the idea of an ensemble is rooted in the dynamical formulation of quantum theory as well; see [2] for details.

The laws governing the system dynamics are deterministic (and thus, respect the principle of causality), in the sense that, if we specify the state of the system at a given instant, for example, $(\psi(\mathbf{x}, 0), \mathbf{x}(0)) = (\psi_0(\mathbf{x}), \mathbf{x}_0)$, then its state at any other (past/future) instant is uniquely determined. However, though such an initial condition is exactly specifiable in theory, in an actual experimental setup it is never known precisely[4]. We therefore have to deal with an ensemble of similarly prepared systems, rather than with a single system. Like in classical theory, an ensemble in quantum theory is also to be understood as a set of fictitious copies of the physical system of interest in a given situation, each independently evolving according to the deterministic laws of motion. Each element (i.e. an individual physical system) from the ensemble is a representative of the possible (wavelike and particlelike) behaviour of the physical system of interest. Also, each element of the ensemble corresponds to a unique initial condition belonging to the set of initial conditions

[4] This could be due to various reasons, such as the coarseness of the instruments, the presence of random background noise and the other external influences.

potentially available to the physical system of interest. This way, it differs from the other elements of the ensemble.

In spite of this fact, the elements from a given ensemble are to be considered as identical to each other in the sense that the wave aspect of each element obeys the same Schrödinger equation[5] and its particle aspect satisfies the same equation of motion, (2.47). Thus, not only the dynamical formulation of quantum theory, but also the inevitable experimental limitations (pertaining to, in particular, the process of *state preparations* (see, e.g., [11]) pave the way for the idea of an ensemble (of identically prepared identical physical systems) – with the additional need to reconcile theory with the world of experience. The Schrödinger equation can then be interpreted as the law of evolution of statistical *ensembles* of similarly prepared (i.e. same ψ) systems.

In quantum experiments, an ensemble may refer to a sequence of identical dynamical processes taking place at different times. Alternatively, it may represent a set of identical processes occurring 'all at once,' that is, simultaneously (see, e.g., [12]). In either of these cases, an individual dynamical process works as a trial contributing to the probability density distribution to be observed as the outcome of a given experiment. A large number of trials leads to the density distribution over the states (e.g., 'positions' or 'spins') of the identical physical systems. In quantum theory, the motion of the elements from a given ensemble is made compatible with the experimental observations of the density distributions by following the Born rule: the probability that an element from the ensemble lies between the points $\mathbf{x}$ and $\mathbf{x} + d\mathbf{x}$ at time t is given by $R^2(\mathbf{x}, t)d^3x$, where $R^2 = |\psi|^2$.

2.6 The "Free Particle" Wave Function, Operators

Let us now turn to the description of the state of motion of a system in terms of its 'wave function.' The simplest dynamical problem is that of the motion of a free particle, that is, of a particle whose motion is not influenced by any forces. With $\mathbf{F} = 0$, it follows from Newton's second law [see Equation (2.1)] that a free particle moves in space with a definite momentum ($\mathbf{p}$). In quantum theory, following de Broglie's hypothesis, the momentum and energy of such a particle are expressed in terms of the wave vector ($\mathbf{k}$) and the wave frequency as follows (see [3]):

$$\mathbf{p} = \hbar\mathbf{k} \quad \text{and} \quad E = \hbar\omega. \tag{2.48}$$

With $\lambda = 2\pi/|\mathbf{k}|$, it follows that the so-called de Broglie wavelength of the wave can be given by[6]

$$\lambda = h/|\mathbf{p}| = h/(2mE)^{1/2}. \tag{2.49}$$

5 By "same Schrödinger equation" we mean that the Hamiltonian and the boundary conditions for all the elements from a given ensemble are the same.

6 For the particle moving under the influence of a classical force, $\mathbf{F} = -\nabla V$, Equation (2.49) generalizes to $\lambda = h/|\mathbf{p}| = h/[2m(E - V)]^{1/2}$.

The wave associated with such a particle is nothing but a plane wave, which can, in general, be described in terms of the 'stationary state' wave function

$$\psi(\mathbf{x}, t) = A\, e^{i(\mathbf{k}\cdot\mathbf{x}-\omega_{\mathbf{k}} t)}, \quad \omega_{\mathbf{k}} = \hbar \mathbf{k}^2/2m \tag{2.50}$$

where A is a constant.

It can be easily checked that

$$-i\hbar\nabla\psi = \mathbf{p}\psi \quad \text{and} \quad i\hbar\frac{\partial}{\partial t}\psi = E\psi, \tag{2.51}$$

where the expression $-i\hbar\nabla$ is called the momentum operator (denoted by $\hat{\mathbf{p}}$) and the expression $i\hbar\frac{\partial}{\partial t}$ is called the energy operator.

The above forms of the momentum and the energy operators are w.r.t. the space-time representation, and hence these can be seen as *differential* operators, whereas the position operator in the space-time representation is simply a *multiplicative* operator. An example of the eigenfunction of the position operator is the Dirac delta function $\delta(x-a)$. With $|\mathbf{k}| = k$, the one-dimensional delta function is often introduced as the nonconverging integral (see, e.g., [3] for details):

$$\delta(x-a) = (2\pi)^{-1} \int_{-\infty}^{-\infty} \exp\left[ik(x-a)\right] dk. \tag{2.52}$$

One of the properties of the Dirac delta function is

$$f(x)\,\delta(x-a) = f(a)\,\delta(x-a), \tag{2.53}$$

from which it follows that $x\,\delta(x-a) = a\,\delta(x-a)$, which can be understood as an eigenvalue equation with 'x' as the position operator, '$\delta(x-a)$' as the position eigenfunction, and 'a' as the position eigenvalue.

2.7 The Uncertainty Principle, Expectation Value

According to Heisenberg's uncertainty principle, the product of the standard deviation uncertainties Δx and Δp with respect to the determination of the positions and the momenta of similarly prepared physical systems is on the order of $\hbar$. More precisely,

$$(\Delta x)\cdot(\Delta p) \geq \frac{1}{2}\hbar. \tag{2.54}$$

The position and the momentum uncertainties, respectively Δx and Δp, are defined in terms of their respective operators' probability weighted averages (also called the expectation values), $\langle\hat{x}\rangle$ and $\langle\hat{p}\rangle$, as follows:

$$\begin{aligned}(\Delta x)^2 &= \langle(x-\langle\hat{x}\rangle)^2\rangle \\ (\Delta p)^2 &= \langle(p-\langle\hat{p}\rangle)^2\rangle.\end{aligned} \tag{2.55}$$

The expectation value of a quantum operator $\hat{A}$ – associated with a classical dynamical variable A of a physical system – is given with respect to a certain state of the physical system. If we describe such a state in terms of $\psi(x, t)$, then $\langle \hat{A(t)} \rangle$ is defined as follows:

$$\langle \hat{A(t)} \rangle = \frac{\int_{-\infty}^{-\infty} \psi^*(x, t) \hat{A}\, \psi(x, t)\, dx}{\int_{-\infty}^{-\infty} \psi^*(x, t)\, \psi(x, t)\, dx}, \tag{2.56}$$

where ψ^* is the complex function conjugate to ψ. The product $\psi^* \psi$ is often expressed as $|\psi|^2$, that is, the square of the modulus of the wave function. For the wave function given by Equation (2.50), the integral $\int_{\text{all space}} |\psi|^2\, d^3x$ does not exist. In other words, the plane wave function is *not* square integrable.

2.8 Wave Packets

Continuing from the previous section, we can also say that the plane wave of Equation (2.50) – associated with a free particle of definite momentum – is completely *delocalized*.[7] It describes the idealized situation in which the particle is free and hence has a definite momentum, but, with $|\psi|^2 = A^2$ (following from (2.50)), the probability that the particle lies within an infinitesimal volume element around any point in space is the same! Following (2.54), this implies an infinite uncertainty in the position and no uncertainty in the momentum of the particle [3].

The way out of the difficulty arising due to such an extreme situation is to give up the idea of 'a free particle having a precisely defined momentum and existing anywhere in the universe' and consider the particle as *confined* to a certain region of space. In such a case the state of motion of the particle cannot be described by a single plane wave function. Instead, we are required to represent the particle by a so-called *wave packet*, which is an outcome of the superposition of a large number of plane wave functions with different wave numbers (**k**). Of course, this way, the particle will no longer have a definite momentum. For such a particle, both the position and the momentum will be *finitely* uncertain. The normalized 'minimum uncertainty' wave packet (satisfying the equality in 2.54) turns out to have the following form (see [8] for details):

$$\psi(x) = [2\pi(\Delta x)^2]^{-1/4} \cdot \exp\left[-\frac{(x - \langle x \rangle)^2}{4(\Delta x)^2} + \frac{i\langle p \rangle x}{\hbar}\right]. \tag{2.57}$$

The concept of a wave packet is useful in the theoretical analysis of the diffraction and interference phenomena. In the following, we briefly discuss the usefulness of the "wave packet" description in the quantum mechanical analysis of Fraunhofer

7 This goes hand in hand with the fact that the plane wave function is *not* an eigenfunction of $\hat{x}$, the position operator.

diffraction (FD) at a single slit. The "FD" experimental setup involves a source, a slit, and a screen. The source can be modelled as a provider of an ensemble of identically prepared identical systems incident on the slit. Here, recalling the discussion from Section 2.5, we consider an individual system from such an ensemble as both a particle *and* a wave by following the de Broglie–Bohm (dBB) formulation of quantum theory. The slit represents both a *confinement* and thus an *allowed* range of the initial positions of the individual systems. Such a position confinement makes a "wave packet" description of the quantum state of the individual systems appropriate. The screen represents the end point of the journey of the identical systems, on which the diffraction intensity distribution of the "particle" aspect of the systems is to be observed.

We assume that at $t = t_{initial} = 0$ the slit produces such a wave packet (with Gaussian profile) along the x-axis. We also assume that the wave packet has initial width ρ and is centred at X. Also, we have $\langle \hat{\mathbf{p}} \rangle = \hbar \mathbf{k}$, so that the centre of the packet has initial (group) velocity $\mathbf{u} = \hbar\mathbf{k}/m$ and $u = |\mathbf{u}|$ is the group speed. Following [13], the initial form of the wave function, that is, $\psi(x, 0)$, coming out from the slit is given by

$$\psi(x,0) = \pi^{-1/4}\rho^{-1/2}\exp\left[-\frac{(x-X)^2}{2\rho^2} + \frac{im}{\hbar}(ux)\right]. \tag{2.58}$$

The free Schrödinger evolution [i.e. with $V = 0$ in Equation (2.35)] of (2.58) leads to the following function at time t:

$$\psi(x,t) = \pi^{-1/4}\rho_t^{-1/2}\exp\left[-\frac{(x-X-ut)^2}{2\rho\rho_t} + \frac{im}{\hbar}(u)\left(x - \frac{ut}{2}\right)\right], \tag{2.59}$$

with $\rho_t = \rho\left(1 + \frac{i\hbar t}{m\rho^2}\right)$.

We can interpret the preceding wave function in accordance with the dBB "pilot wave" interpretation and thus view it as a wave guiding the motion of the particle aspect of the individual ensemble elements from the slit to the screen. For analyzing the phenomenon of single-slit diffraction with respect to the dBB formulation, it is convenient to express $\psi(x, t)$ in the polar form [Equation (2.37)], which provides the amplitude (R) and the phase (S) of the wave function. For details of such an analysis, we refer the reader to [2]. Here we give only the elements of the analysis.

As mentioned at the end of Section 2.5, the square of the modulus of the wave function amplitude ($|\psi|^2 = R^2$) gives the probability density, to be matched with the intensity distribution observed on the screen. Using the gradient of the phase function, that is, ∇S, we can write an equation for $\dot{\mathbf{x}}$ [see Equation (2.47)] and obtain the particle trajectories by integrating it. The amplitude function R also leads to the expression for Q, the quantum potential energy, Equation (2.42), in terms of which we can express a "quantum force" acting on the particles, $\mathbf{F} = -\nabla Q$

[see Equation (2.46)], during their flight from the slit to the screen. This way, not only can we restore the notion of force in the quantum paradigm but can also view the occurrence of the phenomenon of single-slit diffraction as taking place in a "causal" manner. It is interesting to notice that even though $V = 0$ and therefore, "classically" speaking, the motion of the particles should be considered as "free," quantum mechanically it is not! The quantum force arising from the quantum potential energy leads to a distribution of the particle trajectories, which is symmetric about the central position ($x = 0$) on the screen – and this is how the observed intensity distribution could be explained by following the pilot wave interpretation.

As a simple extension to the "classically free" (i.e. $V = 0$) case, we can consider a situation in which the motion of the particles, from the slit to the screen, is influenced by a potential for which the corresponding potential energy depends linearly on the particle position, that is, it can be given by $V = -\mathbf{K} \cdot \mathbf{x}$[8]. In this case, the Schrödinger evolution of the initial wave function (2.59) turns out to be (see [13]):

$$\psi(x,t) = \pi^{-1/4}\rho_t^{-1/2} \times \exp\left[-\frac{\left(x - X - ut + \frac{Kt^2}{2m}\right)^2}{2\rho\rho_t} + \frac{im}{\hbar}\left(u - \frac{Kt}{m}\right)\left(x - \frac{ut}{2}\right) - \frac{iK^2t^3}{6m\hbar}\right]. \tag{2.60}$$

Again, we can express the preceding wave function in polar form and obtain the plot of R^2, the probability density. Just like the case for $V = 0$, in this case as well, we can obtain the expressions of trajectory equation and quantum potential energy. The quantum force in this case has contributions from both the classical as well as the quantum potential energy:

$$\mathbf{F}_{total} = -\nabla(Q + V) = -\nabla Q + \mathbf{K}. \tag{2.61}$$

This time we anticipate that the plot would not be symmetric about the central position ($x = 0$) on the screen. Rather, it would show that the influence of the potential breaks the symmetry of the distribution of intensity and the particle trajectories in this case would tend to bend more either toward the plus or toward the minus side of the x-axis, depending on the direction of **K**, that is, the force due to the potential. In Chapter 10, we give plots of R^2, Q, and the particle trajectories with respect to both the "$V = 0$" and the "$V = -Kx$" conditions, for the case of two-slit interference.

[8] An example of such a situation is the particle motion under gravity ($V = -mgx$), where g is the acceleration due to gravity.

References

[1] Goldstein, H., Poole, C. and Safko, J. (2002). *Classical Mechanics*. 3rd ed., Pearson Education, Inc.
[2] Holland, P. (1993). *The Quantum Theory of Motion*. Cambridge University Press.
[3] Bransden, B. and Joachain, C. (2000). *Quantum Mechanics*. 2nd ed., Pearson Education Ltd.
[4] Athalye, V. and Haven E. (2023). Causal viewpoint and ensemble interpretation: From physics to the social sciences. *Philosophical Transactions of The Royal Society: A*, A381: 20220279.
[5] Penrose, R. (2004). *The Road to Reality*. Jonathan Cape, London.
[6] Schleich, W., Greenberger, D., Kobeb, D. and Scully, M. (2013). Schrödinger equation revisited. *Proceedings of the National Academy of Sciences*, 110, 5374–5379.
[7] Field, J. (2011). Derivation of the Schrödinger equation from the Hamilton–Jacobi equation in Feynman's path integral formulation of quantum mechanics. *European Journal of Physics*, 32, 63.
[8] Schiff, L. (1955). *Quantum Mechanics*. McGraw-Hill.
[9] Bohm, D. and Hiley, B. (1993). *The Undivided Universe*. Routledge, Taylor & Francis Group.
[10] Hall, M. and Reginatto, M. (2016). *Ensembles on Configuration Space: Classical, Quantum and Beyond*. Springer: Switzerland.
[11] Peres, A. (2002). *Quantum Theory: Concepts and Methods*. Kluwer Academic Publishers.
[12] Athalye V. S., Roy S. S. and Mahesh, T. S. (2011). Investigation of the Leggett–Garg inequality for precessing nuclear spins. *Physical Review Letters*, 107, 130402.
[13] von Bloh, K. and Athalye, V. 2022. The Double-Slit Experiment in a Uniform Potential. *Wolfram Demonstrations Project*. https://demonstrations.wolfram.com/TheDoubleSlitExperimentInAUniformPotential/.

3
Elements of Economics

Chapter 2 gave a flavour of some of the relatively technical descriptions of key concepts in classical and quantum mechanics. The concept of ensemble within both classical and quantum mechanics is of great importance, especially in light of how that concept can be used in the mapping exercise from quantum physics to social science. Chapter 3, now coming, is a much less technical chapter which essentially moves in what we could call a graded way: that is, we provide a brief overview of the vast edifice of economics knowledge. We discuss the idea of the 'observable' and the key issue of theory falsification. We also discuss newer developments in economic science, such as experimental economics and computational and neuro-economics. We also consider the connection of neuro-economics with quantum mechanics. The chapter does end with the key notion of 'information' to which we will come back in the book, especially in relation to the quantum-like approach.

3.1 The Many Flavours of Economics: A Brief Introduction

A lot of key advances in economic science in the twentieth and twenty-first centuries will share one common characteristic: they are the result of a rigorous (and often mathematical) development. However, economics, as it was developing in the eighteenth and nineteenth centuries showed a somewhat different evolution. In the eighteenth century, economics developed along the pathway of philosophy and logic. A key logician who played an important role in constructing the foundation of what is later called 'economic science' is Stanley Jevons [1]. In essence, one could claim that Jevons may well be one of the first intellectuals who envisioned the representative agent economy we alluded to earlier. He can also be considered as an important figure in advancing the idea of maximizing utility. However, the recognition that there is room in intellectual discourse for the existence of 'Laws of Economics' had occurred already one century earlier. In fact, the famous invisible hand, which we discussed earlier, was proposed by Adam Smith [2] in the later years of the eighteenth century.

Economics, even in the nineteenth century was probably not yet all that driven by an input of data, as would have been the case with physics. This leads into the debate, even present today, about the difference between normative and what we could call "scientifically based" economics. Normative economics is mostly theoretical and may (or may not) be verified with data. What we called "scientifically based" economics provides for, again, theoretical and rigorous ideas, but they are explicitly falsifiable. A good example of that distinction is the expected utility framework. No matter the type of probability involved, this theory is formulated with a set of axioms. Some of those axioms have been found to be falsifiable by observing typical decision-making behaviour of economic actors.

The history of economics as a developing science, shows that after the nineteenth century, it started acquiring different flavours. Pure theory remained remotely connected to predictability. However, then there was theory which became data driven and led to the development of econometrics. The input of econometrics in macroeconomics has been of pivotal importance. Maybe it is reasonable to claim that there is a dichotomy in economics between a focus on predictability (and in some sense "usefulness") or a focus on a specific theoretical development. If predictability is of utmost importance, then one can make the argument that the often cited reproach that exact scientists make against the unreasonable assumptions used in economics is weak. A high emphasis on predictability can lead to the exclamation: "the model works but we do not really know why." Hence, the assumptions used in the model are of secondary importance. Theoretical economics, on the contrary, will put a very large lens on investigating and discussing the theoretical prowess of assumptions in a model.

The foregoing discussion so far may well raise eyebrows. An exact scientist may never be satisfied with a good predictive model without understanding the accuracy of the model itself. In some sense, this is not only a scientific stance, but it also may be culturally laden. The common argument is this: the "economic" lab is full of incredibly complex interlaced connections leading to very difficult-to-understand causal structures. To make abstraction from those highly complicated intricacies is a prerequisite to make sure that one can construct a workable model. But the price one pays is that many assumptions become very much difficult to fathom in terms of their connection with reality. In contrast to this, the exact sciences may just have a much better handle on the relation between assumptions and the connection to reality of those assumptions.

The reality of the mosaic of economic approaches is much more complicated than what we have been saying so far. For instance, an example of a branch of economics which has neither a theoretically rigorous emphasis nor predictability embedded is Marxian economics. Austrian economics is another branch of economics which shirks away from too much formalism. Other more general approaches to economics will use rigour and predictability in various doses. Some examples are public economics and game theory. Public economics is less mathematically sophisticated than most of game theory.

3.2 The Twentieth Century: Economics Develops at Breakneck Speed

The twentieth century may well be seen as the era, "par excellence," where economics developed as a formal discipline. The contributions from John Maynard Keynes to public economics and especially macroeconomics are of key importance. His treatise entitled "The General Theory of Employment, Interest and Money" (see [3]) is required reading for many economics students specializing in macroeconomics. The contributions of Keynes are many but an important one is the awareness Keynes brings about the belief that economics has tools which can aid in augmenting society's social welfare. Economics is not a tool of solely a capitalist society which is meant to enrich economic actors. Economics has a specific social function. Whilst the invisible hand does many good things for society, certain problems in society such as wage growth and unemployment may not be left to the market to resolve. In effect, Keynes's ideas were unmistakably important in governments willing to overspend and incur so-called deficits. Roosevelt's "New Deal" may well have been inspired by his writings. Finally, Keynes also did have important thoughts about how prices form. This is still, even now, a difficult topic to handle in the financial economics literature. No agreed views exist in the economics community on what precisely the ingredients are in the price formation process of assets and services.

Another important economist who shortly preceded Keynes was Alfred Marshall. Whilst Keynes made hugely important contributions to macroeconomics, also known under the name of "Keynesian economics," Marshall established the foundations of microeconomic theory [4]. The basic concepts of demand and supply and many allied ideas such as the different types of surpluses and the concept of elasticity were all the making of Alfred Marshall. Keynes, in his work, used aggregate demand and supply functions to explain his approach to the workings of the macroeconomy. For instance, the aggregate supply function exhibits a high gradient at what is often called "potential output." At that level of output in the economy, we obtain the "natural rate of unemployment," which is essentially a level of full employment.

The twentieth century is replete with other major contributors to the discipline. We can mention, amongst others, Gérard Debreu and Kenneth Arrow for their key contributions to the theory of General Equilibrium. We surely also can mention

Paul Samuelson and more recently, international economist Paul Krugman. All were Nobel prize winners. There are many others we could name, and our book is not the place where we should expand on any of the key contributions from those towering names.

3.3 The Observable in Twentieth-Century Economics

Here comes a salient question: to what degree does the 'observable' become important in 20th-century economics and beyond? From the foregoing, one may infer that it is not misplaced to state that data has an important role to play in the running of economic models. It is also proper to claim that, as we can deduce from what has been said earlier, there are areas in economics where data and thus the "observable" play a much lesser role. Here is an important quote from Nobel prize winner Paul Samuelson: "the pure theory of consumer's behaviour, when its empirical meaning is finally distilled from it, turns out to be one simple hypothesis on price and quantity behaviour" (see [5]). We note the word 'hypothesis,' which then refers to demand and supply.

3.4 Economics and Theory Falsification

The juxtaposition of the exact sciences with economics, is fraught with subtleties and difficult arguments. Noneconomists may see economics to contribute, at best, as a social engineering discipline. Assume that it were to be the case economics is social engineering (which it is not), then, using the ideas of Popper, one would need to care for the falsification of theories (rather than the confirmation of theories). If we just go back to the famous statement of Samuelson in the former section, that statement he made needs to be falsifiable, rather than confirmed. Popper's idea revolved around the idea of generalizing, inductive statements. If I observe many white cats in nature, I can still not generalize that "all cats are white." However, if I observe a black cat, then I know I have falsified the statement that "all cats are white." Clearly, a lot of statistical techniques and concepts have greatly helped in falsifying and testing theories. This is exactly where econometrics has played such an important role in testing models upon which economic theories can be assessed to be right or wrong. However, this again is simplistic. Econometrics very often helps to calibrate theories rather than falsify them. The observable we know from science is not the same type of observable in economics. To come back to the complex juxtaposition of science with economics, let us look at the following quote (p. 71): "History also shows that the practice of science has not been one of rejecting theories when observation conflicts with theory." (See [6]). So, one may ask, what is the difference then between economics and exact science? One possible answer is that economics has very few "laws." Another possible answer is that in economics we have in effect no constants. Science has many, the Planck constant is an example.

3.5 Constants and Unobservables

The issue of the absence of a constant and the problem of the observable in economics is to some good degree answered by a theory which made huge inroads into financial economics and in effect started a multitrillion market. The theory of financial option pricing which we briefly hinted at in the first chapter, originally devised by French mathematician Louis Bachelier (Bachelier [7]) was revived in the 1970s by Myron Scholes and Fisher Black in their seminal paper (Black and Scholes [8]). The "constant" in this theory is in effect the risk free rate of interest. It is a time-dependent variable. At a fixed time, its level can be observed by many economic actors as being the same. This key 'constant' also does away with the necessity of having to use un-observables in this theory[1]

We discussed in the first chapter "utility," which clearly is un-observable. Samuelson in effect developed revealed preference in the 1930s (see Samuelson [9]) to bypass this problem of un-observability. In his 1948 paper, which also deals with revealed preference, he mentions at the end of the paper (p. 251): "The whole theory of consumer's behavior can thus be based upon operationally meaningful foundations in terms of revealed preference." (see [10]). We note the word "operationally." Bruce Caldwell in his paper [11] says in a footnote (p. 757) that Samuelson in developing his revealed preference theory, was influenced by physics Nobelist Percy Bridgman who developed operationalism, "the view that a phenomenon is defined through the operations by which it is measured."

Finally, even the absence of realistic assumptions in some economic theory models was already supported by Popper. His ideas on so-called zero methods explain it clearly (see Redman [6] (p. 75)): "constructing a model on the assumption of complete rationality (and perhaps on the assumption of the possession of complete information)." Complete rationality and complete information are most probably not realistic assumptions, but models can be constructed on those assumptions.

3.6 Behavioural and Experimental Economics

Whilst economics in the twentieth century (mostly up to the 1980s) revolved around macro- and microeconomics and econometrics in a varying balance of using data and theory, the 1990s and beyond started really embracing novel approaches

[1] Although one can claim that risk neutral probabilities, which form the core of the option valuation protocol, are unobservable.

to economics more specifically in the form of behavioural economics; experimental economics; neuro-economics and computational economics. Most of those areas also generated their own journals and it can be claimed that some of those journals entered the array of mainstream economics.

Behavioural economics makes active use of ideas from psychology and is closely allied to so-called experimental economics. One of the issues with the running of experiments in the academic social science literature is that it is often based in a lab setting where experiment participants are often university students. This is surely not always the case but there are good examples where the use of such a lab context occurs. This issue leads often to the reproach that some of the research results are based on so-called stationary replication which is essentially a replication of a problem setting within a lab setting. As an example, the decision to bet on an object is being repeated multiple times in a lab setting. The connection with reality may be difficult to find. Above all, the problem of context is a difficult one to control for in a lab setting. What is the influence of context in a lab setting as opposed to real life? As is remarked in work by George Loewenstein [12] a key issue with lab-setting experiments is the reward for participating. As he aptly remarks (p. F31) rewards can trigger "the desire to behave in an appropriate fashion, conform to the expectations of the experimenter." However, it is quite shallow to infer judgment on those two fields from this vantage point only. In effect, a Nobel prize was awarded to both Daniel Kahneman (with Amos Tversky) and Vernon Smith for precisely their groundbreaking work in using psychology (in the case of Kahneman) in decision-making under uncertainty; and lab experiments (in the case of Smith) to better understand market mechanisms.

An excellent question posed in the paper by Ana Santos [13] is as follows: are both core principles of classical economics: efficiency and rationality seen to be altered with the arrival of behavioural and experimental economics? According to Santos (p. 708), Vernon Smith would exclude from experimental economics individual decision-making experiments. According to Santos, experimental economics would deal more with market performance. However, she also quotes Camerer and Loewenstein (p. 3) [14] in saying that they "consider instead that experimental economics is defined by the laboratory methods, as applied to both individual choice and market behaviour."

Where probably behavioural economics starts differing from classical economics is really by its modelling of the idiosyncratic attitudes of decision-makers which often provoke decision-making paradoxes. In that regard, one could claim that efficiency and rationality are much more tempered notions in behavioural economics. It is no secret that theoretical economics has continued to keep the formal edifice of utility maximizing, but it has surely accommodated some of the paradoxical decision-making behaviors with alternative, but still formal models.

3.7 Computational Economics

Another recent orientation is so-called computational economics. This approach has tackled very difficult financial puzzles and fundamental assumptions upon which a lot of financial economics theory rests. One important assumption, namely, the existence of a martingale which essentially says that the future asset price, at time $t+1$, given all the information obtained up to time t, is just simply the asset price at time t. This assumption says no prediction is possible. The existence of a martingale is closely related to the efficient market hypothesis. Essentially asset prices follow a random walk induced by white noise. The introduction of different types of traders, like noise and informed traders has rendered the modelling of the financial system much richer, and, it has also tackled very fundamental issues. Clearly, behavioral economics has helped tremendously in modelling the economy with a variety of agents (such as the noise/informed traders dichotomy). Computational economics will take it maybe one step further and in some sense, may also distance itself from the "variety of traders" argument. Recent work by, for instance, Gabaix and Koijen [15] shows that, even uninformed traders (i.e. noise traders) have an effect on market capitalization. If one buys an index (therefore any distinction between informed or noise traders is superfluous), the market capitalization is seriously affected. This remark also goes deeper toward a very fundamental idea that a lot of academic finance will uphold, namely, the efficiency of markets. Any students in a computational economics course will learn about the autocorrelation of returns even on quite long time horizons. This has been observed many times and it has an important implication. The rationale for the existence of martingale given this nonzero observed autocorrelation is not anymore warranted.

Many mainstream arguments in computational and financial economics revolve around the existence of a so-called fundamental value of an asset. A simple example taught in undergraduate finance courses is the so-called Gordon dividend model, which simply states that the price of a stock is the present value of its dividends. Gabaix and Kojien [15], (p. 2) argue that under a dividend model, in an efficient market setting, the effect of an investor buying additional worth of stock will not influence the aggregate value of the market. Another example comes from basic option theory. A call option has an intrinsic value which is the maximum of two arguments: the difference of the asset price and the strike price or zero. Let us keep it simple and assume this intrinsic value (which is not subjective at all) occurs at the end of the option contract. The option price, at time 0 could be written as the expectation of the discounted, using a risk-adjusted parameter, future cash flow (which is the intrinsic value). The probabilistic assumption one makes in order to calculate the expectation is important. Although option pricing does provide for fundamental values, there are caveats to be observed[2] which makes it

[2] The assumption of constant volatility is an example.

come about that the correspondence of fundamental value and observed value will not be perfect. The situation with the fundamental value of other assets is maybe a deeper problem. Fundamental values of stocks can be obtained, to some degree, with models like the capital asset pricing (CAPM) model (see Lintner [16]) which establish a relationship between risk and return.

A key, unrealistic assumption is that investors would be worried only about the mean and variance of the returns of the portfolio. Hence, the CAPM model was extended in subsequent work, notably by Nobelist Robert Merton [17]. The APT model (see [18]) and the five-factor model of Fama and French (see [19]) are other examples of models.

3.8 Economics and Neuroscience: Neuro-economics

As we have mentioned already, lab data has become an integral part in research of the "new economics." Besides lab data, field data is obviously another possibility. More recently, medical data can be used also in decision-making scenarios. More precisely, in work by Frydman et al. [20] tests are devised to measure specific brain processes involved in specific financial decision-making scenarios. This work actually can connect a specific area of the brain (the ventromedial prefrontal cortex) with the calculation a decision-maker makes when deciding to sell a stock. The authors' work also indicates that there is a connection between the activity of another precise area in the brain, the ventral striatum, and the capital gains obtained from selling a stock. Clearly, such a result is precise and steps away from any normative modelling. Very few assumptions are made (or need making) and the position one takes toward what is the observable is quite precise. We seem to be far away from Popper's zero method and even operationalism.

However, there are also even more subtle issues which are revealed from this work. We mentioned earlier that Kahneman's contributions to using psychology in economics were very important.[3] Kahneman with coauthor Amos Tversky are probably best known for so-called prospect theory (see Kahneman and Tverksy [21]) which, in a nutshell, shows that the expected utility framework does not work. The theory predicts that so-called risk-aversive and risk-loving behavior occur when, respectively, people lose money and gain money (on, for instance, stock transactions). This theory is shown actually to also not work in the setting of the capital gains experiment Frydman et al. conducted. This is an interesting result. Specific processes in the brain, when identified, can explain certain decision-making processes. Neuro-economics has made important contributions to our understanding of some of the peculiarities of human decision-making.

[3] We note that their contributions are surely steering away from the zero method position.

How is it then that using the formalism of quantum mechanics could make any contribution? Can we identify any quantum processes in the brain?

Here is an example which may illustrate possible quantum processes in the brain. The psychiatric illness, known as bipolar disorder, is often treated with lithium. In recent work, Hadi Zadeh-Haghighi and Christoph Simon [22] argue that lithium isotopes affect so-called radical pairs. The authors report that the chemical literature indicates that chemical reactions can be altered by nuclear spins. They further indicate that radical pairs need to be created simultaneously which would then require that electron spins are entangled. In some sense a cognitive process influenced by the chemical reaction with lithium is itself dependent on the simultaneity of radical pairs which requires entangled spins.

Other authors, like Matthew Fisher [23] argue that the hydrolysis reaction of pyrophosphate (p. 596) "requires a full quantum treatment of the molecular rotations." The author continues (p. 600): "To be functionally relevant in the brain, the dynamics and quantum entanglement of the phosphorus nuclear spins must be capable of modulating the excitability and signalling of neurons," which we take as a working definition of quantum cognition.

We have remarked in other work (see [24]) that the rationale for using Hilbert space to register brain states leads one to use the idea of the quantum brain (see [25]).

3.9 The Value of Information

There are two basic stances taken in academic finance, with regards to the value information has in the price formation process. One approach is really what we could call 'information based'. This approach discounts to a large degree all noise information. In other words, noise traders (who basically are assumed to trade on the basis of information which is regarded as having "no value") cannot really influence the price. However, the value of private information (which may have immense value) is of key importance then in the process of price formation. The role of private information has been modelled. In recent work, Detemple and Rindisbacher [26] show that the Sharpe ratio (a measure of the excess of the risky return over the risk-free rate per unit of risk), previously driven by public information can also be augmented with an additional component which is reflective of the existence of private information. This additional component Detemple and Rindisbacher call the "private information price of risk," which in their words (p. 190) "represents the incremental price of risk assessed when private information becomes available."

The other basic stance does not actually discount noise information. In other words, even with noise information, noise traders can influence the price.

The efficient market hypothesis does assume that market participants are fully informed and hence one could claim that all public information is fully sufficient to explain price changes. This is an idealization which has been shown to be quite (too) strong. The debate on how public information can explain price changes is not a new one. Kenneth French and Richard Roll did find many years ago, that public information on the NYSE explains only to a small extent stock returns (see [27]). An important point to note is that a random walk (with white noise) can be either the outcome of fully informed traders (as per the efficient market hypothesis), or the outcome of exactly the opposite: traders who are not informed at all.

Carl Hopman [28] shows that if orders have the same magnitude of effect on price, that is, when buy (sell) orders are effectuated, the log price is amended with $\pm\lambda$ (for instance, a sell order drops the log price with λ). As Hopman states (p. 52), if we assume $\pm\lambda$ is randomly occurring in the same way as flipping a coin, then with the central limit theorem the log price change $p_t - p_0$ will follow a random walk. This model is clearly devoid of any information.

References

[1] Jevons, S. (1871). *The Theory of Political Economy*. McMillan.
[2] Smith, A. (1776). *An Inquiry Into The Nature And Causes of Wealth of Nations*. Random House.
[3] Keynes, J. M. (1997). *The General Theory of Employment, Interest, and Money*. Prometheus books.
[4] Marshall, A. (1920). *Principles of Economics*. MacMillan Press.
[5] Samuelson, P. (1965). *The Foundations of Economic Analysis*. Atheneum. The quote is contained in the paper (p. 758) by Caldwell, B. (2013). Of positivism and the history of economic thought. *Southern Economic Journal*, 79 (4), 753–767.
[6] Redman, D. (1994). Karl Popper's theory of science and econometrics: the rise and decline of social engineering. *Journal of Economic Issues*, 28(1), 67–99.
[7] Bachelier, L. (1900). Théorie de la spéculation. *Annales de l'Ecole Normale Supérieure*, 17, 21–86.
[8] Black, F.; Scholes, M. (1973). The pricing of options and corporate liabilities. *Journal of Political Economy*, 81, 637–659.
[9] Samuelson, P. (1938). A note on the pure theory of consumer's behaviour. *Economica*, V, 61–71.
[10] Samuelson, P. (1948). Consumption theory in terms of revealed preference. *Economica*, 15(60), 243–253.
[11] Caldwell, B. (2013). Of positivism and the history of economic thought. *Southern Economic Journal*, 79 (4), 753–767.
[12] Loewenstein, G. (1999) Experimental economics from the vantage-point of behavioural economics. *The Economic Journal*, 109(453), F25–F34.
[13] Santos, A. (2011). Behavioural and experimental economics: are they really transforming economics? *Cambridge Journal of Economics*, 35(4), 705–728.
[14] Camerer, C., Loewenstein, G. (2004). Behavioral economics: past, present, future, pp. 3–51). In Camerer, C., Loewenstein, G. and Rabin, M. (Eds.). *Advances in Behavioural Economics*. Princeton, NJ: Princeton University Press.

[15] Gabaix, X.; Koijen, R. (2021). In search for the origins of financial fluctuations: the inelastic market hypothesis. *NBER*, Working Paper 28967.
[16] Lintner, J. (1965) There valuation of risk assets and the selection of risky investments in stock portfolios and capital budgets. *Review of Economics and Statistics*, 47(1), 13–37.
[17] Merton, R. (1973). An inter temporal capital asset pricing model. *Econometrica* 41:5, 867–887.
[18] Ross, S. (1976). The arbitrage pricing theory of capital asset pricing. *Journal of Economic Theory*, 13(3), 341–360.
[19] Fama, E., French, K. (2015). A five-factor asset pricing model. *Journal of Financial Economics*, 116(1), 1–22.
[20] Frydman, C., Barberis, N., Camerer, C. and Bossaerts, P. and Rangel A. (2014). Using neural data to test a theory of investor behavior: an application to realization utility. *Journal of Finance*, 59 (2), 907–946.
[21] Kahneman, D.; Tversky, A. (1979) Prospect theory: an analysis of decision under risk. *Econometrica* 47, 263–291.
[22] Zadeh-Haghighi, H., Simon, C. (2021). Entangled radicals may explain lithium effects on hyperactivity. *Scientific Reports*, 11: 12121.
[23] Fisher, M. (2015) Quantum cognition: the possibility of processing with nuclear spins in the brain. *Annals of Physics* 362, 593–602.
[24] Haven, E. and Khrennikov, A. (2013). *Quantum Social Science*. Cambridge University Press.
[25] Hameroff, S. (1994) Quantum computing in brain microtubules? The Penrose–Hameroff Birch OR model of consciousness. *Philosophical Transactions of the Royal Society – London A*, 1–28.
[26] Detemple, J., Rindisbacher, M. (2016). The private information price of risk. pp. 190–213. In *The Handbook of Post-Crisis Financial Modelling* (Haven, E. et al., eds.). Palgrave MacMillan – Nature Publishers.
[27] French, K., Roll, R. (1986). Stock return variances: the arrival of information and the reaction of traders. *Journal of Financial Economics*, 17, 5–26.
[28] Hopman, C. (2007). Do supply and demand drive stock prices? *Quantitative Finance*, 7(1), 37–53.

4

Some Selected Financial Models

Chapter 4 focuses on some of the important underpinnings of asset pricing and the notion of information will be carefully assessed in the various models we will discuss here. A key contribution that the mapping exercise of quantum physics into economics and finance can make, is to provide as precise a foundation as possible on how to model information in an asset pricing context. A price process is highly dependent on information. However, whilst price formation is actually rather independent of information, we will see that price discovery does have a dependence. The role of information is well captured in a well-known theoretical contribution in finance: the Kyle model, which is discussed here. The Glosten and Milgrom model de-emphasizes the role of information in asset price formation. In this chapter we also consider two other key ideas used in finance – arbitrage taking and the description of dynamic asset price behavior. We round off the chapter with a suggestion on how we could extend the Rosenow model on price generation with one of the hallmark characteristics of Bohmian mechanics, namely the quantum potential.

4.1 Introduction

Although, in a purely economic sense, we can use the supply and demand framework to determine equilibrium prices, there are several mechanisms in economics and finance which will underpin the gestation toward this equilibrium. In other words, the equilibrium is the final result of a process, which needs formalizing. It is useful, in view of the modelling of the double-slit experiment in economics and finance, that we provide for some background on how each of those mechanisms function and how they interlock with each other. The mechanisms will be documented in the following sections, notably when we discuss the Kyle model and the Glosten and Milgrom model.

4.1.1 The Limit Order Book

A key ingredient is the so-called Limit Order Book (LOB) which can be defined, as per Glosten[1], (p. 1129) "as a publicly visible screen providing bids and offers, each of which specify a price and a quantity." Foucault et al. [2] report that Jain [3] shows that in the late 1990s close to half of the world stock markets are using pure limit order books. However, as has been aptly remarked by J.P. Bouchaud et al. [4] (p. 335) "... the visible LOB is only the tip of the iceberg, beneath which lies a large amount of latent liquidity." A key argument made in [4] is that (p. 19) "the instantaneous traded volume is much smaller than the underlying 'true' supply and demand waiting to be matched."

4.1.2 Back to Price Impact

An important cause of the issue that trade volumes can be much smaller than supply and demand volumes, relates to the fact that trades are often broken up so as to reduce as much as possible the impact large orders could have on prices. This relates to the idea of the so-called price impact function which can be defined in general (see Huberman and Stanzl [5]) as to include both temporáry and permanent price impacts. Such impact can be modelled as being time (in)dependent. We note that the price impact function is discussed in Section 4.6 when we elaborate on the Rosenow model. Huberman and Stanzl [5] also define liquidity (a concept we will use below) as the first derivative of the price impact function (toward time).

4.1.3 Price Discovery

Price impact is dependent on the orientation adopted in terms of whether one believes that price follows a discovery process: that is, when fast spreading of information (via for instance, a well-run financial exchange) leads to fast price updating. Another way of looking at price is via the price formation lens, which we discuss in what follows. Essentially, price formation does not depend on information, whilst price discovery does. As Biais et al. [6] indicate, in price discovery all investors can have *equal* access to information. When we discussed, earlier on, the theory of rational expectations, with its rational decision-makers who face the same information, we were there (in some sense) implicitly leaning in close to the idea of price discovery. There is important research that has indicated that large price movements may simply not be (sufficiently) reflective of information. In work by Cutler, Poterba and Summers[7] the authors come to the conclusion that (p. 9) "Our results suggest the difficulty of explaining as much as half of the variance in aggregate stock prices on the basis of publicly available news bearing on fundamental values." A recent study by Biondo, Pluchino, Rapisarda and Helbing [8] (p. 11) comes to the conclusion "that standard trading strategies and their algorithms based on the past history of the time series.... on a large temporal scale perform on average not better than the purely random strategy."

4.1.4 Price Formation as Opposed to Price Discovery

Whilst we discussed price discovery earlier, there is another price process – which Bouchaud et al. [4] call price formation. This process is rather independent of information, and is tied to the act of trading proper. Given some of the preceding findings, the assumption of access to common information, albeit not wrong (given we have access to free public information via databases), still may indicate that the value of this common information is probably low. This, of course, excludes the importance of private information. In the next section we consider one of the most well-known models in financial theory. The Kyle model really carefully identifies the role of information within price setting. Highly novel concepts like market depth were introduced via that model.

4.2 The Kyle Model

The Kyle model asks some very fundamental questions on how fast private information enters prices and what the contribution of noise traders may be in setting the level of liquidity of the market. Those are key research questions which were answered in a rigorous way in the Kyle [9] model. For our purposes, it is important to have background knowledge on the importance of the accumulation and impact of both types of information. A key idea in the Kyle model is the so-called market depth which is defined as [9] (p. 1317): "Market depth is proportional to the amount of noise trading and inversely proportional to the amount of private information (in the sense of an error variance) which has not yet been incorporated into prices." Theorem 1 (p. 1319) is central in the model, and the market depth (being the inverse of the Kyle Lambda, λ) indicates that the level of market liquidity, in the words of Kyle [9] (p. 1319): "is proportional to a ratio of the amount of noise trading to the amount of private information the informed trader is expected to have." The lambda is then defined as:

$$\lambda = 2\sqrt{\frac{\Sigma_0}{\sigma_u^2}}, \tag{4.1}$$

where Σ_0 is the variance of the ex-post liquidation value of the risky asset and σ_u^2 is the variance of quantity traded by noise traders and $1/\lambda$ is a measurement of the depth of the market. A very important relation is given by how the expected profit of the insider (i.e. the market participant who has private information) relates to both Σ_0 and σ_u^2 :

$$\frac{1}{2}\sqrt{\Sigma_0\sigma_u^2}. \tag{4.2}$$

Hence, both noise and private information contribute to the expected profit of the private information trader. Bouchaud et al. [4] say (p. 293): "the conditional expectation of Alice's[1] gain increases with the amount of private information and with the overall liquidity of the market." Theorem 2 in Kyle [9] yields a relation for λ which now features an intensity parameter β, and

$$\lambda_n = \frac{\beta_n \Sigma_n}{\sigma_u^2}, \tag{4.3}$$

where n denotes the nth auction. β is defined as per Kyle [9] (p. 1323): "the intensity with which the insider trades on the basis of his private observation." Hence, a high intensity will reduce market depth and reduce liquidity and therefore will augment the cost of trading. In Kyle and Obizhaeva [10] an illiquidity measure is introduced and the measure is essentially proportional to the volatility of returns (in so-called business time).[2]

The intuition here is that illiquidity is spurned on by high return volatility. According to Kyle and Obizhaeva [10], the measure is also connectable to Emanuel Derman's notion of market temperature. Essentially the higher the liquidity of a market, the higher the market temperature[3]. Derman [11] defines this temperature (p. 291) as being proportional to "the product of calendar time volatility and the square root of trading frequency."

Kyle also finds that private information becomes gradually incorporated into prices. A key conclusion is that (Kyle [9], p. 1326): "increasing the amount of noise trading increases market depth proportionally...and leaves the informativeness of prices unchanged."[4]

For our purposes, depth is related to cost of trading and therefore can correspond to how trading impacts price. With high depth (and low λ) the price impact of volume is small. We can possibly also invoke here an argument by Pástor and Stambaugh [12] where the authors mention that liquidation of assets is less costly when liquidity is high as opposed to when it is low. Finally, we can claim that with high depth, the instantaneous traded volume will be close to the 'true' supply and demand, i.e. price impact is low.

[1] Alice is defined as the informed trader.

[2] See [10] for more information on the background of the illiquidity measure.

[3] See Kyle and Obizhaeva [10], p. 1358 for a more detailed discussion.

[4] We note that in the paper by Kyle [9], besides the sequential auctions (we briefly hinted at these earlier when defining n in λ_n) there also is a development on continuous auctions. We do not discuss this issue here.

The next model we consider does connect the bid-ask spread with the inverse of the market depth parameter as introduced by the Kyle model we just discussed.

4.3 The Glosten and Milgrom Model

The Glosten and Milgrom model defines the transaction prices to form a martingale, and this, relative to *both* specialist and public information[5]. This clearly puts little value on information in the price process. In Bouchaud et al. [4], (p. 301) the bid and ask prices are defined as (per the market maker):

$$a = p_0 + s/2 \tag{4.4}$$

$$b = p_0 - s/2, \tag{4.5}$$

where p_0 is current price and s is bid-ask spread. As Bouchaud et al. [4] (p. 302) show[6], s can be approximated by $2\phi\sigma$, where ϕ is the fraction of informed trades and σ is the width of the informed signal[7]. In some sense, $2\phi\sigma$ could be interpreted as a measure of temperature in the way of Derman [11] as mentioned earlier, where trading frequency could be approximated by ϕ.

As remarked in Back and Baruch [14], the bid-ask spread can be seen in relation to the λ parameter we discussed above. Theorem 3 in Back and Glosten [14] (p. 447) relates the variables as follows:

$$\lambda = \lim_{\delta \to 0} \frac{a_\delta(p) - p}{\delta} = \lim_{\delta \to 0} \frac{p - b_\delta(p)}{\delta}, \tag{4.6}$$

where δ is trade size and where $a_\delta(p)$ and $b_\delta(p)$ are respectively the ask and bid prices.

Glosten and Milgrom in [13] provide for some very useful examples to highlight aspects of their approach. Essentially, the examples show how bid and ask prices can be formed, and especially how insider traders affect information diffusion. See [13] , p. 91. As a flavour from those examples, the authors show that given a proportion of insiders (relative to the full trading population) and a prior probability on the future value of a financial asset, bid and ask prices can be determined. Those prior probabilities can also be updated following trades. The example also shows that when the number of trades is sufficient to lead to assimilation of information, there is a good indication of how the bid-ask spread narrows with an increasing proportion of insiders.

[5] This is proposition 2, p. 82 in Glosten and Milgrom [13].

[6] Uninformed traders must have information signals with amplitude w wider than σ (width of informed signal).

[7] Which is related to the volatility of the fundamental price.

4.4 Asset Prices: Trading and Trajectories

Before we tackle the Section 4.5, price formation, we want in this section to make a small detour to devote some time to elementary ideas behind some of the stochastics (read: formalization of randomness) used in finance. The distinction between discrete and continuous time also is important. In the financial world, trading can occur on different times scales, essentially ranging from discrete time to an idealization, which is known as continuous time. Trading cannot really be continuous as such (although it is nearly so especially with algorithmic trading) but this idealization forms the basic starting assumption of many useful models in finance. If that is the case, then the question which needs asking is the one which David Kreps [15] eloquently formulated: "when and whether discrete-time models of economic phenomena 'converge' in economic terms to their mathematical limits." In what follows here, we shall show some examples illustrating discrete and continuous time trading with associated trajectories.

4.4.1 Discrete Time - Arbitrage Taking

Arbitrage is an essential concept in basic finance and can be essentially defined as making a riskless profit.[8] Here is an example which illustrates arbitrage taking on a discrete time scale, and it also exemplifies that physics ideas may actually, almost naturally, be embedded in financial strategies.

Kirill Ilinski [16] shows the following interesting strategy. Given a cash amount, we want to buy shares in two different ways:

i) we have cash at time t_i: we want to buy shares at time t_{i+1}: where $\frac{\exp(r_0\Delta)}{S_{i+1}}$ is the number of shares you can buy at time t_{i+1} at price S_{i+1}. The accumulated cash at time t_{i+1} with start value 1 at time t_i is given by $\exp(r_0\Delta)$
ii) Alternatively – one can buy shares now at t_i at price S_i and at time t_{i+1}: $\frac{\exp(r_1\Delta)}{S_i}$

Consider then the following two situations:

$$\exp(r_0\Delta)S_{i+1}^{-1} > \exp(r_1\Delta)S_i^{-1} \tag{4.7}$$

and the opposite:

$$\exp(r_1\Delta)S_i^{-1} > \exp(r_0\Delta)S_{i+1}^{-1}. \tag{4.8}$$

A nonarbitrage condition could be:

$$\exp(r_1\Delta)S_i^{-1}\exp(-r_0\Delta)S_{i+1} + \exp(r_0\Delta)S_{i+1}^{-1}\exp(-r_1\Delta)S_i - 2 = 0 \tag{4.9}$$

[8] The martingale measure can be used to discount expected asset payoffs within a so called finite state space. For an excellent paper on unrealized arbitrage opportunities, see Zimper [17]

In Chapter 2, we first encountered the Lagrangian and the idea of the action function. The Lagrangian is $\mathcal{L} = T - V$, where T is kinetic energy and V is potential energy. The least action idea

$$\delta\mathcal{A} = \int_{t_o}^{t_1} \mathcal{L}(x, \dot{x})dt = 0, \tag{4.10}$$

is central to both classical and quantum mechanical formulations. The relationship with the preceding and the Euler–Lagrange equation:

$$\frac{d}{dt}\left(\frac{\partial \mathcal{L}}{\partial \dot{x}}\right) = \frac{\partial \mathcal{L}}{\partial x}, \tag{4.11}$$

is one from which we can recover

$$F = ma. \tag{4.12}$$

An action can be formulated as (see [16]):

$$\mathcal{A} = \sum_{i=-\infty}^{\infty} \alpha_i(\exp(r_1\Delta)S_i^{-1}\exp(-r_0\Delta)S_{i+1} + \exp(r_0\Delta)S_{i+1}^{-1}\exp(-r_1\Delta)S_i - 2), \tag{4.13}$$

where α_i is dependent on variance. This action is to be minimized. We note that the minimization of this action seems to have also financial meaning. The question becomes: if that is the case, what is the consequence of such minimization? Is there financial meaning we can assign to the use of the Euler–Lagrange equation?

4.4.2 Continuous Time: the Brownian Motion on Stock Prices

The Brownian motion is a well-known mathematical object in physics. From a formal point of view we could treat it as a so-called stochastic differential equation (SDE), where the word "stochastic" refers essentially to the random component of this SDE. A general form of motion is

$$dx = a(x, t)dt + b(x, t)dW, \tag{4.14}$$

where dW is a Wiener process, which in discrete form can be written as

$$\Delta W = \varepsilon\sqrt{\Delta t}, \tag{4.15}$$

with $\varepsilon \sim N(0, 1)$, which means the random variable is drawn from a Gaussian density with mean 0 and variance 1.

In finance, following on from the early work of Louis Bachelier: set x as a stock price S, and $a(x, t) = \mu S$; where μ is an expected return; $b(x, t) = \sigma S$, where σ is the variance:

$$dS = \mu S dt + \sigma S dW. \tag{4.16}$$

Note that Bachelier did opt for a wrong version of this Brownian motion:

$$dS = \mu dt + \sigma dW, \tag{4.17}$$

where stock prices can be negative (which cannot be the case). As is pointed out by Damiano Brigo and Fabio Mercurio [18], p. 2, "the geometric Brownian motion is not the only continuous time process that possesses a lognormal marginal distribution and normal log returns."

Finally, the Brownian motion trajectory is continuous but nowhere differentiable. For more on Brownian motion, please do consult [19]. For more on the background of Louis Bachelier's work, please see [20] and see also the chapter by Mark Davis in [21].

4.4.3 Continuous Time: Brownian Motion and Volatility

We can use Brownian motion for emulating stochastic behavior of other variables, like volatility. Suppose we have the stochastic differential equation for volatility, σ:

$$d\sigma = \alpha(\sigma)\, dt + \beta(\sigma)\, dX. \tag{4.18}$$

Paul Wilmott in [22] shows that one can estimate $\alpha(\sigma)$ using the Fokker–Planck PDE approach. Under the assumption of a steady-state distribution, using the Fokker–Planck PDE:

$$\frac{\partial P}{\partial t} - \frac{1}{2}\frac{\partial^2 P}{\partial \sigma^2}\beta^2 + \frac{\partial P}{\partial \sigma}(\alpha) = 0, \tag{4.19}$$

one can state

$$\frac{1}{2}\frac{\partial^2 P_\infty}{\partial \sigma^2}\beta^2 = \frac{\partial P_\infty}{\partial \sigma}(\alpha) \tag{4.20}$$

and integrating once (toward σ):

$$\alpha P_\infty = \frac{1}{2}\frac{\partial P_\infty}{\partial \sigma}\beta^2, \tag{4.21}$$

from which we derive

$$\alpha = \frac{1}{2P_\infty}\frac{\partial P_\infty}{\partial \sigma}\beta^2, \tag{4.22}$$

where the constant is assumed zero. Here we might wonder, what would be the potential function when considering the preceding volatility model?

4.4.4 Brownian Motion: Bonds and Forward Rates

Brownian motion can also be invoked within the context of bond pricing. We can express the bond price, $B(t, T)$ on a \$1 face value (and continuous compounding) as $B(t, T) = \exp(-R(T - t))$, where R is the long yield. We can also express $B(t, T)$

in terms of so-called forward rates: $F(t, T, U)$, where forward rate contract is made up at time t, starts operating at time T and expires U, $U > T > t$. We can express forward rates in terms of bond prices:

$$F(t, T + \Delta) = \frac{\log B(t, T) - \log B(t, T + \Delta)}{\Delta}. \tag{4.23}$$

If we are interested in finding an expression for the infinitesimal change in the forward rate, we can here, precisely, invoke the use of Brownian motion on bonds. From the requirement we want to look for:

$$dF(t, T + \Delta) = \frac{d \log B(t, T) - d \log B(t, T + \Delta)}{\Delta}, \tag{4.24}$$

one can show that when we introduce a geometric Brownian motion on the bond price, B:

$$dB = rBdt + \sigma BdW \tag{4.25}$$

then we can write dF in terms of diffusion and drift terms. The calibration of the volatility parameter, σ, will be essential in the final expression for dF.

4.4.5 Brownian Motion and Information

A key concept in finance is the so-called martingale, to which we alluded earlier in this book. The martingale is defined as:

$$E(S_{t+u} | I_t) = S_t, \tag{4.26}$$

where I_t is the information set taken up to time t. The conditional expectation of the future asset price, S_{t+u}, is thus equal to S_t if there exists a martingale. Hence, information in this setting is of virtually no value. A martingale implies there are no trends:

$$E(S_{t+u} | I_t) - E(S_t | I_t) = 0. \tag{4.27}$$

A Brownian motion certainly has a trend, but this can be remedied. Finally, for continuous martingales, S_t:

$$E(S_t^2) < \infty. \tag{4.28}$$

This is the square integrability condition. This means there is variability AND there should be no jumps. From a financial point of view, specifically option pricing, jumps in a Brownian motion trajectory would pose issues as to the uniqueness of the option price.

4.5 Toward Pricing

When mapping physics concepts into social science, we must be continuously aware of the pitfalls that identical words can mean very different things depending on where those words are placed. A good example is the use of microscopic, mesoscopic, and macroscopic levels of price formation. When discussing how the limit order book may only be a part of the price process and latent liquidity explains quite a chunk of the missing information leading up to what one observes as prices (at the macroscopic level), we are in some sense implying different scales at which prices occur. Those scales, here microscopic/mesoscopic and macroscopic, are widely different from what those identical words mean in physics. As an example, the microscopic scale in physics refers to the atomic scale where time is measured in units which are fully foreign to what humans measure their time in. In the next section we consider how macroscopic prices form, that is, the prices we observe in the market.

4.5.1 Price Revisions

We subscribe to the interesting idea that Bouchaud et al. in [4] have in defining price revisions as follows:

$$p_{t+dt} \rightarrow p_t + \beta_t d\xi + dW, \tag{4.29}$$

where idiosyncratic noise is dW and $d\xi$ is what Bouchaud et al. call public news. The parameter β_t is particularly interesting and measures the sensitivity of an agent to news.[9]

4.5.2 States and preparation

We now think of two states, where in one state, the price $p_t > p_{t+dt}$ and in the other state, we have $p_t < p_{t+dt}$. We must have identically prepared agents who face the same context. Prima facie, we could link this important assumption to the idea of so-called homogeneous expectations, which is a well-known concept in basic finance. The word here is used to indicate that market participants expect the market to be in some format, which is homogeneous across agents in that all the statistical information agents have is the same across all participants. We do want to make it explicit, though, that identical preparedness is not akin to the financial notion of expectations which are homogeneous. Let us consider the parameter D, which in Bouchaud et al. [4], p. 342, is defined as follows:

$$D = \frac{1}{2}(\Sigma^2 + \sigma^2 S[.]). \tag{4.30}$$

[9] Please see Bouchaud et al. [4], p.340 and also equation 18.2.

where Σ^2 is variance sourced from idiosyncratic *noise*; whilst $S[.]$ refers to variance coming from what we like to call the (in)homogeneity of information. We juxtapose the concept of inhomogeneity of information with the expectation of homogeneity of information, which we said earlier is not the same as identical preparation. We hope it is now clear that we mean the information is simply not the same across participants. Preferences from participants need to be expressed based upon the existence of this information and this is reflected in the parameter β. As in Bouchaud et al. [4], p. 342, $S[.] = 0$ for $\beta = 1$. We can say that $S[.]$ is not zero for β when there is over- or underreaction (i.e. $\beta \neq 1$). We would want to argue that the diffusion coefficient D is dependent on the sensitivity to information or, one could say, the degree of information the ensemble has. We may want to argue that if $\beta = 1$, we could claim that the ensemble has maximal information. Alternatively, we could claim that under $\beta = 1$ we have the homogeneity of expectations across agents in the ensemble. This could be seen as akin to the requirement of identically prepared agents facing an identical context. In effect, $\beta = 1$ could be interpreted as indicating the presence of high levels of private information. In other words, the over- and underreaction to information could be seen as the consequence of not having sufficient information to avoid such reactions in the first place.

The treatment by Bouchaud et al. in [4] shows that under a high level of D, liquidity is large and in some sense one could argue that the presence of noise creates liquidity in the market. In the extreme, a market where much private information prevails could not be called liquid. Noisy markets will be markets where volatility must be high as opposed to markets where there is a marked absence of noise.

4.6 The Rosenow Model

An important question is how the time evolution of an instantaneous return relates to so-called order imbalance. Such a question, together with the fundamental work of Hopman, which we cited earlier (see Section 3.9), can aid us in providing a fairly simple base upon which we can graft our approach in the sequel of the book (and also in further work). Bouchaud et al. in [23] discuss a financial equivalent to the fluctuation-dissipation theorem. We shall follow here the work of Rosenow in [24] where the financial analogue is also explained, albeit in a different way from the approach that Bouchaud et al. follow.

In [24], the analogue to the fluctuation-dissipation theorem in physics is proposed. The analogue to the (time-dependent) variance of positions of particles is in finance seen to map onto the (time-dependent) variance of returns. The analogue of the temperature and particle mobility in physics is captured by the so-called friction, which Rosenow equates to market liquidity. Bouchaud et al. in [23] make an important distinction between the types of diffusive behavior occurring when one discusses market-making versus market-taking behavior. The latter can be associated with the group of informed traders, whilst the former relates to providers of liquidity. It needs to be made clear what is meant by liquidity. In essence, a market is liquid when one can trade a commodity/service in the quantity and at the time of one's choice. Providers of liquidity (who can be both sellers and buyers) basically trade on short horizons so that the market can buy and sell at any point in time for any quantity. Hence, there is continuity in trading. Bouchaud et al. make an important argument in that they advance that market providers can cause what is called subdiffusive behavior. The mean squared displacement is a power law of time (with an exponent smaller than unity), whilst so-called informed traders trigger a power law with an exponent on time in excess of unity.

The former model which we termed the Hopman model (see Section 3.9), was a good introduction on how one can generate dynamic log price changes. In the Rosenow model, the number of financial variables used in dynamic log price generation is now increased.

4.6.1 Financial Background

To provide for the essentials of the model we need to provide first some financial background.

- the model incorporates order imbalance. In basic finance, we encounter sell or buy order imbalances. Sell order imbalances occur when a big stock position is closed out. This will lead to price drops. A buy order imbalance often occurs with the occurrence of valuable private information and it can lead to a price increase. Buy orders are in excess of sell orders in the latter case, whilst the reverse is the case in the former.
- The Rosenow model argues that the transaction price lies between bid price (the highest price buyer is willing to buy at) and the ask price (the lowest price a seller will accept).
- A key point in the Rosenow model is the financial analogue of the physics-based proportional relation between the fluctuation of position and temperature and the so-called particle mobility.

4.6.2 Formal Setup of the Model

We now list the key formal steps in the formulation of the Rosenow model.[10] A basic start revolves around defining the log stock price change ΔS simply as

$$\Delta S = \ln S(t) - \ln S(t - t') \tag{4.31}$$

[10] We do follow the Rosenow [24] paper, closely (pp. 2–3). We did change some of the notation.

The instantaneous return, R_I, is defined as

$$R_I = \frac{d \ln S(t)}{dt}\tau, \tag{4.32}$$

where τ is the average time interval between trades and hence $\frac{1}{\tau}$ is then an indicator of trade frequency. If we sum instantaneous returns over a time interval Δt, one can write

$$R_{\Delta t} = \frac{1}{\tau}\int_{t-\Delta t}^{t} R_I dt. \tag{4.33}$$

The preceding equation is a very straightforward way of writing returns over a period of time Δt. Such a setup can be found in many other models such as, for instance, the Heath Jarrow Morton model on (forward) interest rates (see Heath et al. [25]).

The instantaneous order imbalance, Or_I, is defined using the volume imbalance, which we denote in capital letters as OR:[11]

$$Or_I = \frac{dOR}{dt}\tau. \tag{4.34}$$

The last ingredient one needs to set up the model is what Rosenow [24] defines as the slope of the so-called price impact function. We shall quote directly Rosenow [24] (p. 3): "The price impact function describes the functional relation between the expectation value of $R_{\Delta t}$ conditioned on the volume imbalance and the volume balance itself." The slope of the price impact function, χ, is then defined using the basic idea of a derivative:

$$\chi = \lim_{\Delta OR \to 0} \frac{E(R_{\Delta t})}{\Delta OR}. \tag{4.35}$$

4.6.3 The Langevin Equation Expressing the Dynamics of R_I

$$\frac{dR_I}{dt}\tau = \frac{-1}{\chi}R_I(t) - \kappa R_I^3(t) + Or_I(t) + \xi(t); \tag{4.36}$$

The preceding equation (Langevin equation) can be found in [24], p. 2. κ is a free parameter and $\xi(t)$ is white noise. As Rosenow remarks, one can interpret $\frac{1}{\chi}$ as an indicator of liquidity. This is intuitive. A high slope of the price impact function would indicate low market liquidity.

The Langevin equation indicates that the effect of a high-order imbalance, $Or_I(t)$, is having a direct effect on the time dynamics of the instantaneous returns.

11 In the Rosenow paper [24], a buy order imbalance is assumed.

The market liquidity has the opposite reaction. Assume there is a high-order imbalance with a highly liquid market. Then in that case the high level of $\frac{-1}{\chi}$ will offset the direct effect of the order imbalance on the instantaneous return dynamic. Hence, the dynamics are tempered. However, if there were to be a high-order imbalance with low liquidity, then the effect of that order imbalance will weigh heavily on the instantaneous return dynamics and the dynamics are now heightened.

We note again that high liquidity, that is, $\frac{1}{\chi}$ is high and therefore χ is low, translates well, intuitively, in low standard deviation of transaction prices.

4.6.4 *Random Liquidity and Noise Contribution: How Information is an Implicit Parameter in the Langevin Equation*

The white noise term, $\xi(t)$, in the Langevin equation is defined as containing information which is *not* contained in the order imbalance parameter. Rosenow also makes the remark that the liquidity level, if random, can lead to multiplicative noise in the Langevin equation. We can argue that this can lead to sub- (or superdiffusive behavior).

4.6.5 *Diffusion Constant*

The financial equivalent of the diffusion coefficient is now

$$D = 2\chi^2 \frac{1}{\tau} \Delta t, \tag{4.37}$$

where $\frac{1}{\tau}$ is the trading frequency and it is akin to what in physics would be temperature.

4.6.6 *First Steps Toward Understanding Sources of Information*

From Newton's second law, $m.\frac{dv}{dt} = F$, we can describe the motion of a Brownian particle as being influenced by (i) frictional force, $-\delta v$, and a random force, denoted as F_r. One can thus write the Langevin equation for a Brownian particle as:

$$m\frac{dv}{dt} = -\delta v + F_r. \tag{4.38}$$

Let us now do a naive exercise. Consider the augmented Newton's law, which now includes the "force"[12] associated with the quantum potential which we discussed earlier in the book. This is defined as

$$ma = -\nabla V - \nabla Q. \tag{4.39}$$

[12] Physicists have debated what type of force the negative gradient of the quantum potential portrays.

Now reconsider the Langevin equation we considered earlier:

$$\frac{dR_I}{dt}\tau = \frac{-1}{\chi}R_I(t) - \kappa R_I^3(t) + Or_I(t) + \xi(t). \tag{4.40}$$

The left-hand side of this financial Langevin equation corresponds to "acceleration." Set

$$-\kappa R_I^3(t) + Or_I(t), \tag{4.41}$$

in its physics equivalent, equal to

$$-\delta v = -\nabla V. \tag{4.42}$$

The random force F_r, would then in financial terms correspond to

$$\frac{-1}{\chi}R_I(t) + \xi(t). \tag{4.43}$$

We can assume, as per our earlier discussion, that liquidity is randomized.

We recall the discussion that information not contained in the order imbalance is contained in the random term $\xi(t)$. Can we think of defining

$$\frac{-1}{\chi}R_I(t) + \xi(t) = -\nabla Q? \tag{4.44}$$

If yes, then the information captured by the force driven by the quantum potential could refer to information contained in

$$\frac{-1}{\chi}R_I(t) + \xi(t). \tag{4.45}$$

In Chapter 6, we shall come back to the modelling of information.

References

[1] Glosten, L. R. (1994) Is the electronic open limit order book inevitable? *Journal of Finance*, 49(4),1127–1161.

[2] Foucault, T., Kadan, O. and Kandel, E. (2005). Limit order book as a market for liquidity. *The Review of Financial Studies*, 18(4), 1171–1217.

[3] Jain, P. (2002). Institutional design and liquidity on stock exchanges. *Working paper*. Bloomington, IN: Indiana University.

[4] Bouchaud, J. P., Bonart, J., Donier, J. and Gould, M. (2018). *Trades, Quotes and Prices: Financial Markets under the Microscope*. Cambridge, UK: Cambridge University Press.

[5] Huberman, G., and Stanzl, W. (2004). Price manipulation and quasi-arbitrage. *Econometrica*, 72(4), 1247–1275.

[6] Biais, B., Hillion, P. and Spatt, C. (1999). Price discovery and learning during the pre-opening period in the Paris Bourse. *Journal of Political Economy*, 107(6), 1218–1248.

[7] Cutler, D., Poterba, J. and Summers, L. (1989). What moves stock prices? *Journal of Portfolio Management*, 15(3), 4–12.
[8] Biondo, A. E., Pluchino, A., Rapisarda, A. and Helbing, D. (2013). Are random trading strategies more successful than technical ones? *PLOS One*, 8(7), e68344.
[9] Kyle, A. (1985). Continuous auctions and insider training. *Econometrica*, 53(6), 1315–1335.
[10] Kyle, A., and Obizhaeva, A. (2016). Market microstructure invariance: Empirical hypotheses. *Econometrica*, 84(4), 1345–1404.
[11] Derman, E. (2002). The perception of time, risk and return during periods of speculation. *Quantitative Finance*, 2(4), 282–296.
[12] Pàstor, L., and Stambaugh, R. F. (2003). Liquidity risk and expected stock returns. *Journal of Political Economy*, 111(3), 642–685
[13] Glosten, L., Milgrom, P. (1985). Bid, ask and transaction prices in a specialist market with heterogeneously informed traders. *Journal of Financial Economics*,14, 71–100.
[14] Back, K., Baruch, S. (2004). Information in securities markets: Kyle meets Glosten and Baruch. *Econometrica*, 72(2), 433–465.
[15] Kreps, D. (2019). *The Black-Scholes-Merton Model as an Idealization of Discrete Time Economies*. Cambridge, UK: Cambridge University Press.
[16] Ilinski K. (2001). *Physics of Finance: Gauge Modelling in Non-Equilibrium Pricing*. Hoboken, NJ: J. Wiley & Sons
[17] Zimper, A. (2023). Unrealized arbitrage opportunities in naive equilibria with non-Bayesian belief processes. *Mathematical Social Sciences*, 125, 27–41.
[18] Brigo, D., Mercurio, F. (2000). Discrete time vs continuous time stock price dynamics and implications for option pricing. BANCA IMI, PDG Internal report - Reduced version in *Finance and Stochastics*, 4, 147–159.
[19] Gardiner, C. (1983). *Handbook of Stochastic Methods*. Berlin: Springer.
[20] Davis, M.; Etheridge, A. (2006). *Louis Bachelier's Theory of Speculation: The Origins of Modern Finance.* Princeton, NJ: Princeton University Press.
[21] Haven, E., Molyneux, Ph., Wilson, J., Fedotov, S. and Duygun, M. (2015). *The Handbook of Post Crisis Financial Modelling*. Palgrave MacMillan Handbook Series.
[22] Wilmott, P. (1998). *Derivatives: the Theory and Practice of Financial Engineering*. Hoboken, NJ: J. Wiley & Sons, p. 319.
[23] Bouchaud, J. P., Gefen, Y., Potters, M. and Wyart, M. (2004). Fluctuations and response in financial markets: the subtle nature of 'random' price changes. *Quantitive Finance*, 4, 176–190.
[24] Rosenow, B. (2002). Fluctuations and market friction in financial trading. *International Journal of Modern Physics*, 13(3), 1–7.
[25] Heath, D., Jarrow, R. and Morton, A. (1992). Bond pricing and the term structure of interest rates: a new methodology for contingent claims valuation. *Econometrica*, 60(1), 77–105.

Part II

Quantum-like Modelling

5

Quantum-like Modelling in Economics

In the next two chapters, we provide the reader with some exemplars of work, where the emphasis really lies on the use of the quantum formalism as it is applied to problems in economics (Chapter 5) and finance (Chapter 6). As one will see from Chapter 5, the quantum formalism there centres around accommodating paradoxical outcomes in decision making. Quantum probability offers a response to those decision-making contexts where a consistent violation of the law of total probability occurs. We start out Chapter 5 with elementary ideas such as incompatibility and the conjunction and disjunction effects. We also consider some of the building blocks of mainstream expected utility. In using the quantum-like approach, strong results have been obtained in decision-making applications and the chapter will go into some detail to discuss the so-called QQ equality and will also make mention of the quantum-like version of the well-known Aumann theorem in economics. Chapter 6 considers essentially Bohmian mechanics and its usefulness in finance. We discuss the Hawkins and Frieden approach and we provide for a brief overview of how types of potential functions can be used as devices to distinguish between types of public information when considering asset price return data. We also discuss Bohmian mechanics within the context of a macroscopic fluid bath.

5.1 Introduction

The title of this chapter may confuse readers. The term 'quantum-like' is surely a qualifier in modelling of economic systems which has not yet really entered mainstream research. How could we define it? In this section, we note what Andrei Khrennikov has termed the 'quantum information revolution.' In the abstract of [1], he writes: "Quantum-like modeling has its basis in quantum information theory, and it can be considered one of the fruits of the quantum information revolution." For further discussion, see [1, 2].

First of all, the formalism we aim for is sourced from quantum mechanics and will apply (by virtue of us speaking about economics) to large collective social systems. 'Quantum-like' brings in formalistic approach – that is, quantum probability. However, the rationale why such non-Kolmogorovian counting is appealing needs to be explained!

There is a realistic existence for so-called incompatible events in the macroworld. When we consider *compatible* events, two events can be realized simultaneously and the order of events does not matter. Incompatible events will negate those two conditions. The most well-known example of incompatibility relates straight back to the famous Heisenberg uncertainty principle: position and momentum are quantum-like incompatible observables.

The order of events issue is well recognized in consistently repeated experiments in psychology. Order effects, that is, asking question 'A' first followed by question 'B', may give different results relative to the case where question 'B' is asked first and is then followed by question 'A'. We can identify that related to this incompatibility, there appears an alternative characterization of uncertainty. It is intuitive, but very difficult to rigorously define. There must be a different type of uncertainty related to events which are commutative in nature as opposed to events which do not commute.

This chapter will start laying out in Section 5.2 some (what we hope are) intuitive arguments, why incompatibility is of such great importance in the arguments which attempt to support using the quantum formalism in economics, especially in the foundational field (for economics) of decision-making. We then continue in Section 5.3 with a brief discussion on conjunction and disjunction effects and more advanced decision-making contexts. In Section 5.4 we discuss some alternatives to the expected utility framework, especially how, via accommodating the various paradoxes, one implicitly shows the usefulness of the quantum formalism. Section 5.5 highlights a first major achievement in quantum-like based decision-making, that is, the establishment of the QQ equality. Finally, Section 5.6 considers the second major achievement: the quantum-like version of a very well-known theorem in economic theory: the Aumann theorem.

Of course, we could extend our discussion further into more complex settings. As an example of a much more sophisticated setting: could the Clauser–Horne–Shimony–Holt version of the Bell inequalities find applicability outside of quantum physics? We would, however, claim that some of those specific quantum results may still be open for debate. Nonetheless, at this point in time it is reasonable to advance the argument that most of the quantum-like modelling results, namely those which have received a wide acclaim of acceptability in the mainstream literature, revolve around decision-making. The fallacies of decision-making as documented with cases such as the Linda bank teller problem and also the sequential gambling problem, are very robust examples of fallacies which in many different experimental settings keep on throwing up the same decision-making inconsistency. We expand on this in what follows in more detail.

5.2 Incompatibility to the Rescue in the Conjunction Fallacy

Amos Tversky and Daniel Kahneman in [3] defined and analyzed the Linda bank teller problem. In this problem, Linda is described to experiment participants as having studied philosophy and as being involved in social justice. Then, the same participants are asked what hypothetical description is more likely: (i) she is a bank teller; (ii) she is a bank teller and is a feminist activist. As is intuitively clear, the expectation should be that most participants should pick option (i). In many repetitions of this experiment, a vast majority of experiment participants will choose option (ii) and in fact it is shown that the probability of the event 'bank teller and feminist activist' is in excess of the probability of 'bank teller'. This immediately contradicts basic probability: the probability of the intersection of two events (in other words the conjunction of events) is the product of the probabilities of each event. Since probabilities are contained in [0,1], it is impossible that the choice taken by experiment participants obeys this simple probability rule. The quantum-like modelling community has correctly argued that this very robust result goes completely against classical probability theory. As Emmanuel Pothos and Jerome Busemeyer [4] remark, the conjunction fallacy arises because of interference which arises because of incompatibility. Quantum probability can accommodate this strange result when one introduces the idea of incompatibility. In essence, within the Linda bank teller problem, it is well known that with the aid of quantum probability it is the case there is no equality between the probability of Linda being a feminist and then a bank teller - and the probability of Linda being a bank teller and then being feminist. This nonequality can easily be seen when considering projections of the state vector onto the rays composed of either 'Feminist' or 'Bank teller'. This is very well illustrated in figure 3 (p. 757), in [4]. The projection of the state vector upon the 'Feminist' ray to be followed with a projection onto the 'Bank teller' ray gives a probability which is different from when we do a first projection of the state vector on the 'Bank teller' ray to then project this projection onto the 'Feminist' ray.

Of course, as we remarked before,[1] it is not so obvious how one can now interpret the human response relative to this use of quantum probability. If it relates

[1] Research Pod - Apple Podcast (2021) Quantum society: The logic of decision making, economics and relationships - Andrei Khrennikov and Emmanuel Haven

to incompatibility, then that is a reasonable explanation, but if it does not, then the proposal of using quantum mechanics is one amongst other possibilities (other than classical probability). See Pothos and Busemeyer in [4], p. 758: "since the conjunction fallacy problem is about probabilistic judgment, we might be biased to develop a model for the conjunction fallacy based on some kind of probability theory. What is a probability theory, not constrained in the same way as CPT, which is consistent with the intuitions expressed in the heuristic, similarity-based models for the conjunction fallacy? The most immediate answer is QPT."[2]

In essence, the quantum formalism does not allow for the measurement of the probability of the intersection of two events, precisely because of incompatibility. This has a very important consequence relative to classical probability: the idea of a conditional probability then does not exist, if we use a quantum-like version of incompatibility. This is Boolean logic based. Let us think more: can we as human beings really assess the conjunction of two events? In many instances, we may not be able to do so. In that sense, one could say that classical probability is an idealization, at least in some applications of the social sciences and humanities. This leaves open also another question: namely what is the interpretation of probability – in, for instance, its quantum guise? Shall it be frequentist?

We can maybe emphasize that the quantum formalism has as its starting point the idea that events are subspaces of a vector space and it is precisely this approach which leads to an extension of Boolean logic, that is, quantum logic where commutativity and distributivity of events do not hold.

5.3 Earlier Ideas on the Quantum Formalism in Decision Theory

Amongst the protagonists who provided fundamental contributions of the quantum formalism to decision making, we include Andrei Khrennikov, Jerome Busemeyer, Emmanuel Pothos, and Ehti Dzhafarov, amongst others. In a recent application for a seminar event at the Fields Institute in Toronto[3], the organizers[4] provided for several arguments on the rationale one can bring forward to using the quantum formalism in social science. Here are some:

1. The conjunction effect which we hinted at in the former section (the Linda bank teller problem) is a robust result. Busemeyer et al. in [5] show this as follows. They consider the conjunction of a person who is over 50 and has a heart attack. They introduce a projection operator P_O as the projector representing the event

[2] CPT refers to classical probability theory, whilst QPT refers to quantum probability theory.

[3] This event happened during March 9–11, 2015 and was titled "Quantum probability and the mathematical modelling of decision-making."

[4] These included Andrei Khrennikov, Jerome Busemeyer, Arkady Plotnitsky, Emmanuel Pothos; Ehti Dzhafarov, and Emmanuel Haven.

that a man is 50 or over. The projector for men under 50 is then $P_U = I - P_O$. They also define the projector denoting a man getting a heart attack as P_H. In quantum formalism terms, the probability that a man gets a heart attack is as follows:

$$|P_H\Psi|^2 = |P_H.I.\Psi|^2 = |P_H.(P_O + P_U)\Psi|^2 = |P_H P_O\Psi|^2 + |P_H P_U\Psi|^2 + IT, \tag{5.1}$$

where IT stands for interference term.

As is well known, the interference term will disappear when projectors commute. A nonzero interference term depends on the level of constructive (or destructive) interference, which Busemeyer et al. in [5] and others relegate to the interference of thoughts. Note that the first term, $\|P_H P_O\Psi\|^2$, equals the probability that the man is over 50 and has a heart attack. If the interference is sufficiently negative so that

$$IT < \|P_H P_U\Psi\|^2, \tag{5.2}$$

then a conjunction fallacy occurs:

$$\|P_H\Psi\|^2 < \|P_H P_O\Psi\|^2 \tag{5.3}$$

because of negative interference. As Busemeyer et al. in [5] also indicate, this gives a good starting point for explaining both the conjunction fallacy and the disjunction fallacy.

2. The disjunction effect can be presented as follows. Experiment participants form part of a sequential experiment containing two events. For instance, in event 1, a gamble is presented, and in event 2, participants are asked whether they would want to (i) gamble again; (ii) not gamble again – and that on the basis of (i) having been informed they won in the first gamble; (ii) having been informed they lost in the first gamble; or (iii) having no information of the outcome of the first gamble. As was found first by Tversky and Shafir in [6] and has been reproduced many times over, in a variety of experimental settings, the finding reveals that experiment participants prefer to do event 2 after event 1 whenever they are informed of the outcome of event 1. However, experiment participants prefer not to go on to event 2 when they do not know the outcome of event 1. The Savage expected utility model (see [7]), which embeds subjective probability in the construction of expected utility, has within its axiomatic structure, a well-known axiom – called the 'sure-thing principle'. In the experimental setting we just described, this axiom is violated. In the Tversky and Shafir study, 69% of experiment participants will want to participate in a second gamble, assuming they know they won in gamble 1. Close to 60% of experiment participants will want to gamble a second time if they know they lost in the first gamble. The very strange result which occurs in the experiment is that only 36% of the experiment

participants will go on to event 2 (thus gamble) if they have no information of the outcome of event 1. We will discuss the Savage model a little more in Section 5.4.

Busemeyer and Wang in [8], p. 92, are correct in advancing the ideas that in a Markovian model one expects that "the probability of gambling in the unknown case must be the average of the probabilities of gambling in the two known cases." Busemeyer and Wang in [8] (pp. 91–92) show that in the Markovian setting there are (i) two states of belief on the outcome of the first gamble; and (ii) there are also two states of action ('gamble' or 'not gamble').

As reported by Haven, Khrennikov and Robinson in [9], in the quantum-like approach, Busemeyer and Wang [8] (pp. 91–92) define basis states with the use of the following kets: $|WG\rangle$ (you simultaneously believe you won in the first gamble and you will undertake a second gamble); $|WN\rangle$; (you simultaneously believe you won in the first gamble and will not gamble). Similarly, for a loss in the first gamble, the basis states are $|LG\rangle$; $|LN\rangle$. The corresponding probability amplitudes are $\psi_{WG}; \psi_{WN}; \psi_{LG}; \psi_{LN}$.

Assuming there exists an initial state vector ψ and information is transmitted that one won or lost, then Busemeyer and Wang in [8] propose that a unitary operator U is applied: $U.\psi$. Thus, as an example, if you are informed you lost the gamble, then the initial state is transformed from

$$[\psi_{WG}; \psi_{WN}; \psi_{LG}; \psi_{LN}] \tag{5.4}$$

to

$$[0, 0, \psi_{LG}, \psi_{LN}] \tag{5.5}$$

and an unknown state will be a superposition of the lost and win states.

3. Another contribution to decision-making has been identified with reference to so-called dynamic inconsistency (see Barkan and Busemeyer in [10]), which when applied to the disjunction effect requires one to optimize starting from the end (i.e. the second event) and move toward the first event. In real-world experiments, Barkan and Busemeyer obliged a first play, but gave the option to participants to decide to play or not to play the second gamble. The distinction here is that there is a so-called planned choice which is conditional on winning or losing in the first gamble and a final choice when players have played and are informed of the outcome of the first event (i.e. the gamble). Barkan and Busemeyer found that experiment participants steered away from their plans on the final decision. They found that winning at the first stage decreased the probability of taking the gamble compared to the planned choice (while losing in the first event increased the probability compared to the planned choice). This violates the law of total probability. The disjunction and conjunction effects also violate it.

The law of total probability can be easily illustrated as follows (see Haven and Khrennikov in [11]) (pp. 122–123). Let $a = +1$: "the democrats will win"; $a = -1$: the negation; $a = +1$. An example for the b-variable: "you buy a condominium in midtown Manhattan", $b = -1$: the negation; $b = +1$. The law of total probability is then as follows:

$$\mathbf{P}(b = j) = \mathbf{P}(a = +1)\mathbf{P}(b = j|a = +1) + \mathbf{P}(a = -1)\mathbf{P}(b = j|a = -1), \quad (5.6)$$

where, $j = +1$ or $j = -1$.

4. Another contribution relates to decision making within random information (noisy) environments. Such tasks, called 'signal detection tasks' can also be found back in financial decision making, for instance (see Busemeyer et al. in [12]). In the latter model, the wave function represents a so-called evidence state. In chapter 8 of [13], Busemeyer and Bruza derive a key prediction that provides a critical method to empirically distinguish the two theories. The experiment consists of two conditions: In the choice-confidence condition, the decision-maker makes a choice (signal present or absent) at time t_1 and then rates confidence at time t_2; in the confidence alone condition, the decision-maker simply presses a button at time t_1 (no decision is required) and only provides a confidence rating at time t_2. For both conditions, the focus is on the marginal distribution of confidence ratings that are obtained at time t_2. Confidence is defined as the probability that a signal is present on a '0 = certain target not present' to '100 = certain target is present' scale. The Markov model obeys the Chapman–Kolmogorov equation (which leads to the Fokker–Planck PDE)[5] and predicts no difference between the two conditions. The quantum model predicts that an interference effect is produced by the decision for the choice-confidence condition that makes the distribution different from that of the confidence alone condition.

5.4 Alternatives to Expected Utility

Utility and its associated concept of 'expected utility' is of paramount importance in any field of economics and thereby is of great value in the development of financial theory.[6] The question a noneconomist may ask is, "What is the formal basis of utility?" We briefly discussed utility at the beginning of the book in Section 1.3.1.

[5] Busemeyer et al. in [12] propose the juxtaposition of the Markovian model (essentially a Fokker–Planck PDE with Brownian motion as underlying stochastic movement) and the Schrödinger PDE.

[6] One of the hallmarks of financial theory is financial option pricing theory. We remarked in Subsection 3.5 that there are unobservables in economics (such as utility) and in option pricing theory too there are nonobservables, like risk-neutral probabilities. One of the important issues to observe is that option pricing as a financial theory is probably one of the foremost best examples where utility plays absolutely *no role* whatsoever.

The following bullet points denote simple and intuitive key properties that bundles of consumption goods (those can be services too) should possess. In what follows, we denote with capital letter X the set of consumption bundles[7]; whilst with lowercase x and y we denote the consumption bundles.

- completeness (we can always express a preference of one bundle over another): $\forall x, y \in X : x \succeq y$ or $y \succeq x$ (or both). We note that $x \succeq y$ denotes bundle x is weakly preferred over bundle y)
- reflexivity: $\forall x \in X : x \succeq x$
- transitivity: $\forall x, y, z \in X : x \succeq y$ and $y \succeq z \Longrightarrow x \succeq z$
- the utility function, u: $u : X \to \mathbb{R} : x \succ y \Longleftrightarrow u(x) > u(y)$

There are two additional properties which are needed besides the three first properties just listed and they are (i) continuity and (ii) strong monotonicity. As is nicely described in Hal Varian's classic micro-economics graduate textbook, [14] (p. 95): "if y is strictly preferred to z and if x is a bundle that is close to y, then x must be strictly preferred to z". This refers to the continuity property. The strong monotonicity can be defined in a self-explanatory way as follows: $x \geq y$ and $x \neq y \Longrightarrow x \succ y$; that is, a good is 'good' to own.

Here now follows the key theorem in basic utility theory and it refers to the *existence* of a utility function: $\mathbb{R}^k_+ \to \mathbb{R}$. This function u will exist and be continuous when the first three preceding bullet points with the addition of (i) continuity and (ii) strong monotonicity are satisfied. Any advanced book in basic economic theory will state and prove this key theorem. We note that if X is countable infinite, the same result can be found which establishes a relation between the preferences over objects and the numerical representation (i.e. via the existence of the utility function). However, for an uncountable set X this relation is fraught with difficulties.

At this point in our discourse, we have not yet discussed the term 'expected' in 'expected utility'. Clearly, this is a statistical reference which brings us right into the context of uncertainty. In there, we shall need to contemplate the types of probability we will be employing to represent this uncertainty. In the expected utility literature, probability can assume different 'characters' – it being objective, subjective or a mixture of both. We shall now delve, in a somewhat formal way, into those divisions.

5.4.1 von Neumann–Morgenstern Expected Utility – in a Nutshell

The key trait of this approach resides in the use of preferences relative to *objective* probability measures over a set of outcomes, denoted as O. Essentially (thus discounting discussions again on the (un)-countability of O), one seeks a relation

[7] The set X is finite here.

between the preference, now over probability measures and the existence of a utility function. We can write that the probability measure is as follows:

$$prob : O \to [0,1] : \sum_{o \in O} prob(o) = 1, \tag{5.7}$$

and the preference relation and the utility function is then as follows:

$$prob_1 \succ prob_2 \Leftrightarrow \sum_{o \in O} prob_1(o)u(o) > \sum_{o \in O} prob_2(o)u(o), \tag{5.8}$$

where *prob* is a probability measure (and which is objective).

Hence, with the use of objective probability we obtain the von Neumann–Morgenstern [15] expected utility model. The main paradox related to this model is the so called Allais paradox ([16]), which violates one of the key axioms of the von Neumann–Morgenstern model, namely the so-called independence axiom. That axiom, in words, essentially says that if I prefer one lottery, l over another lottery q, then adding another lottery, h, to both of those lotteries should not alter my original preference. That 'context'-irrelevant axiom is violated in many decision-making problems.

5.4.2 Savage Expected Utility – in a Nutshell

The *subjective* probability approach in expected utility resides squarely in the so-called Savage expected utility model (see Savage [7]). We referred to this model in Section 5.3. The model is in essence sophisticated and we provide for a couple of ideas here which, we hope, should give some taste of the model. We follow here David Kreps' exposition, [17]. Chapter 9 in Kreps' book outlines that the main ingredients in the Savage expected utility framework are

- a set of consequences, C,
- a set of states of the world, S,
- an act – which is a function mapping S into C. The notation $f(c)$ determines the consequence of taking the action f given the occurrence of a state of the world; and
- the set of all acts f, denoted as F.

What do decision makers express preferences over in this theory? They indicate their preferences over acts. The decision maker has a belief over what consequences, they think, might occur. The Savage framework provides for an axiomatic structure which allows the decision-maker to form a value for the expected utility, with the aid of (thus) now subjective probabilities, of acts taken. As we mentioned in Section 5.3, there is an issue with the so-called sure-thing principle, a key axiom in the formal Savage structure. As Edi Karni [18] remarks: (pp. 10–11): "almost from the moment of its inception, the descriptive validity of Savage's model – in particular, the sure-thing principle, which is responsible for the specific functional form of the representation and the separability and linearity in the probabilities – has been questioned. It has repeatedly been shown, in experimental settings that the theory fails systematically to predict subjects' choice. The most severe and remarkable criticism in this regard is due to Ellsberg in [19], who demonstrated using simple thought experiments that individuals display choice patterns that are inconsistent with the existence of beliefs representable by a probability measure."

The Ellsberg paradox is the main inconsistency observed when using the Savage decision-making axiomatics. The combination of both subjective and objective probabilities in an expected utility model leads to the Anscombe–Aumann model [20]. Narrowly connected to the idea of subjective probability is the issue of the distinction between risk and uncertainty. Risk is measurable by objective probability, versus uncertainty, which connects to the idea of so-called Knightian uncertainty [21]) and only measurable via subjective probability.

In decision-making formalisms there are many more paradoxes besides the Linda teller problem. It needs also to be stressed that approaches other than quantum have been introduced, especially within the economic theory literature. In [22] Aerts et al. distinguish four different groups of models and amongst those four, the first group of models follows a non-Kolmogorovian route where nonadditive capacities are used. In the so-called max-min expected utility models (see Gilboa and Schmeidler [23]) multiple priors are considered; whilst in the smooth ambiguity preferences approach (see Klibanoff et al. [24]) decision-makers are assumed to have a prior over a set of priors.

In summary, in the expected utility literature, probability can assume different characters – being objective, subjective, or a mixture of both. In objective probability we obtain the von Neumann–Morgenstern [15] expected utility model. The main paradox related to this model is the well-known Allais paradox ([16]). The subjective probability approach in expected utility resides squarely in the so-called Savage expected utility model (see [7]). The Ellsberg paradox is the main inconsistency observed when using the Savage decision-making axiomatics. The combination of both subjective and objective probabilities in an expected utility model leads to the Anscombe–Aumann model (see [20]). Narrowly connected to the idea of subjective probability is the issue of the distinction between risk and uncertainty. Risk is measurable by objective probability, versus uncertainty, which connects to the idea of so-called Knightian uncertainty [21]) and is only measurable via subjective probability.

The quantum cognition approach has done much work around accommodating the Ellsberg paradox via the quantum probability route. As is argued for by Aerts et al. in [22] a lot of academic work has been performed within the mainstream economics community on also trying to mend the paradox via the usual Kolmogorovian probability approach. Aerts et al. in [22] quote some key work such as Etner et al. [25]; Gilboa and Marinacci [26]; Machina and Siniscalchi [27]; and Ellsberg et al. [28]; Machina [29]. However, the same authors in [22] also identify essentially four alternatives and we cite two of those:

- (i) Choquet expected utility which uses nonadditive capacities, and
- (ii) 'max-min expected utility' and 'alpha-max min expected utility', which are models essentially geared on using priors.

Following Chateauneuf et al. [30], a capacity is essentially defined via the idea that it is a function defined on a Sigma Algebra of subsets of the set of states (that set describes the uncertainty of the decision maker), and which maps into the real numbers. Here is the definition given by Chateauneuf et al. in [30] (Definition 2.1., p. 540): A capacity is

$$v: \mathcal{E} \to \mathbb{R}; E, F \in \mathcal{E}; E \subseteq F \Longrightarrow v(E) \leq v(F); v(\emptyset) = 0; v(S) = 1. \tag{5.9}$$

We note that S is the set of states (representing uncertainty) and $\mathcal{E}$ is the set of events (the subsets of the set of states S).

The important issue to observe is that capacities are generalizations of probability, just as quantum probability is a generalization of Kolmogorovian-based probability. The capacity can be convex or concave (i.e., respectively):

$$v(E \cup F) \geq v(E) + v(F) - v(E \cap F), \tag{5.10}$$

and

$$v(E \cup F) \leq v(E) + v(F) - v(E \cap F)). \tag{5.11}$$

As Chateauneuf et al. remark, a probability distribution is a capacity which is both concave and convex. The well-known Choquet integral is defined relative to v and for a function $f{:}S \to \mathbb{R}$, as follows in [30] (Definition 2.2., p. 540):

$$\sum_{t \in f(S)} t.[v(\{s|f(s) \geq t\}) - v(\{s|f(s) \geq t\})]. \tag{5.12}$$

Capacities are thus nonadditive probabilities, and thus from here are generalizations of the usual Kolmogorov-based probability. The Choquet capacities are also known under the name of belief functions. What can its use be from a decision-maker's perspective? Essentially, it accommodates the decision maker by allowing them to assign an interval of numbers relative to a level of uncertainty. See also [31].

In the max-min expected utility approach, Gilboa and Schmeidler [23] indicate (p. 142) that when there is convexity, that is,

$$v(E \cup F) \geq v(E) + v(F) - v(E \cap F), \tag{5.13}$$

a Choquet integral of a real-value function a with respect to v. They define (p. 142), the core of v consisting of all "finitely additive probability measures that majorize v event-wise." In the max-min expected utility, the priors will refer to the core of v. To put it in simple terms, max-min expected utility is as per Kelsey in [32], (p. 426) predicting that "action a will be preferred to action b *iff* the minimum possible value of the expected utility of a is greater than the minimum expected utility of b."

In recent literature, the notion of informational asymmetry has come up, via the work of Mark Machina [33] and others. As Aerts et al. in [22] correctly remark, the alternatives which have been proposed have surely opened the door toward recognizing that non-Kolmogorvian probability (such as quantum probability) is a viable alternative.

Aerts and Sozzo in [34] also have taken their research much further and have considered *quantum structures* where ideas such as superposition can find a natural home, within a human cognition setting. The literature they have developed around defining quantum entities as so-called concepts has opened up a new area of research in quantum cognition. A kernel finding in that research is that entanglement can be proven to exist in human language. The authors use the von Neumann entropy measure to show this existence. They also argue that such entanglement has a direct relation with reducing uncertainty in language.

5.5 The QQ Equality and Nonatomic Instruments

Related to the issue of incompatibility is the recent innovation that Wang et al. [35] brought forward with the so called QQ equality (i.e. the quantum question equality). In essence, within the setting of at most two measurements, the QQ equality is formulated as follows. Let $p(A, thenB)$ denote the probability that event A happens first and is then followed by the occurrence of event B. Denote $p(\neg A, \neg thenB)$ as the probability that not A happens followed by not B happening. The QQ equality is then the following constraint:

$$p(A, thenB) + p(\neg A, \neg thenB) - p(B, thenA) - p(\neg B, \neg thenA) = 0. \tag{5.14}$$

This equality was verified in the paper by Wang et al. [35]. James Yearsley [36] does remark, though, that the QQ equality is unlikely to be generalizable to measurements with more than two outcomes (or to more than two measurements). However, it is a very strong result which can best be answered by quoting out of the Wang et al. paper [35] as follows (p. 9436): "Applications of both the QQ equality and a wide class of psychological and decision-making tasks.... share the following conceptual bases: (i) Human judgments, such as attitude judgments, are often not simply read out from memory, but rather, they are constructed from the cognitive state for the question at hand; and (ii) drawing a conclusion from one question changes the context and disturbs the cognitive system, which then (iii) affects the answer to the next question, producing order effects, so that (iv) human judgments do not always obey the commutative rule of Boolean logic."

We have already discussed incompatibility in Section 5.2 and the idea of the conjunction fallacy in Section 5.3. As Emmanuel Pothos and Jerome Busemeyer remark in [37] (p. 760) it is both the conjunction fallacy and the question order effects which can be sourced from interference, itself then resulting from incompatibility. They also emphasize that it is the cooccurrence of both the conjunction fallacy and question order effects which (p. 771) "is another unique prediction from quantum probability theory." The QQ equality we just discussed is, in the words of Pothos and Busemeyer [37], nothing else than the difference between the product of projectors (relative to question A first and question B after) $P_B P_A \psi^2$ – and the projectors corresponding to the reverse order, that is, $P_A P_B \psi^2$, which yields the constraint we mentioned earlier. It is important to mention this result, as it does point, very clearly, to an important advantage in using the quantum probability approach. Pothos and Busemeyer in [37], p. 762, say: "the advantage of QPT (quantum probability theory) is the capability of making precise a priori predictions, such as the QQ equality."

Andrei Khrennikov indicates that the QQ equality can be derived within von Neumann measurement theory, but relative to the so-called response replicability effect there is an issue with using that theory.[8]

Very recently, in work by Masanao Ozawa and Andrei Khrennikov [38] the order effect on question type of the structure A-A and A-B-A was investigated via the use of a novel technique called nonatomic instruments. The A-A question type is simply referring to the asking of a question A with an ensuing answer. The repeat of that same question A (without any additional information) is then supposed to give the same answer. This is called A-A replicability. The question type A-B-A is then a question A with an ensuing response; and another question B, with another ensuing response, with the expectation that the experiment participant will answer the same when the equation A is asked again. This is called 'response replicability'. The so-called response replicability effect combines both A-A and A-B-A (or B-A-B).

[8] We refer the interested reader to a very clear discussion of the response replicability effect in Khrennikov [1], pp. 7–9. The discussion on how von Neumann measure theory will be an issue when that effect is taken into account is presented in Khrennikov [1], p. 16 and also [38].

The question order effect is simply saying that the joint probabilities on questions A and B are not the same when A-B is asked as opposed to when B-A is asked. So changing the order of questions can lead to a different outcome. The key point now is that if one combines the question order effect with the response replicability effect, quantum measure techniques will not work, but the use of nonatomic quantum instruments will.[9] Those are strong results which show that the usability of the quantum approach (but now with commuting observables) is much reinforced.

5.6 The Quantum-like Version of the Aumann Theorem

In economics, there is a well-known theorem by Nobelist Robert Aumann [39]. A neat description of that theorem in words is provided by Hild et al. in [40] as follows: "In 'Agreeing to Disagree', Robert Aumann proves that a group of agents who once agreed about the probability of some proposition for which their current probabilities are common knowledge must still agree, even if those probabilities reflect disparate observations."

We note the type of probability which seems to be implicit here. Common knowledge is information I have on somebody else, and that somebody else has also the same information on me. A common priors probability distribution means essentially that the same probability distributions are assumed on a forthcoming event, which is a very accepted assumption in economics and it will require objective probabilities. How realistic are common priors amongst decision making? If any decision maker has private information for instance, the assumption of common priors will not hold. Efficient markets, on the other hand, could easily assume common priors. Efficient markets could also easily assume common knowledge. A key underlying assumption, however, is that decision makers must be rational. In fact, it has been remarked that rationality implies the assumption of a common prior (see Morris [41]). The Boolean algebra comes into being again: decision-makers can express probabilities over the intersection of events. It needs to be emphasized that the fallacies in decision-making we mentioned in other sections earlier, are very robust and show that such Boolean algebra is an idealization.

Andrei Khrennikov [42] presents a quantum-like analog of the Aumann theorem we cited earlier. The key qualifier is that the possible state of the world is now a mixed quantum state. The quantum-like version negates the theorem – namely quantum agents can agree to disagree (even with common knowledge). The kernel argument, allies with what we have also seen earlier, when we considered the Linda bank teller problem or the QQ equality: in the quantum-like approach we can make the argument that, within decision making, the processing of information

[9] Such instruments relate to so-called positive super operators.

happens via quantum logical rules. The interference term will measure the degree of disagreement in the quantum-like version of the Aumann theorem.

In a nutshell, we can develop the arguments of his paper as follows. Let ρ be a density operator and ρ_C be a density operator updated via common knowledge. Also, an agent updates a prior state ρ, such that (p. 96 [42])

$$\rho \to \rho_{out} \equiv \frac{\rho_C}{Tr\rho_C}, \tag{5.15}$$

where C denotes common knowledge. Another update is

$$\rho \to \overline{\rho_{out}} \equiv \frac{\overline{\rho_C}}{Tr\overline{\rho_C}} \tag{5.16}$$

As indicated in the paper, each update refers to measurements of different operators. The key point resides in that

$$\rho_{out} = \overline{\rho_{out}} + \delta \tag{5.17}$$

and

$$\delta = \frac{1}{Tr\rho_C} \sum_{i \neq j} P_i \rho_C P_j, \tag{5.18}$$

where $P_{i,j}$ are projectors. The equation

$$\rho_{out} = \overline{\rho_{out}} + \delta \tag{5.19}$$

is a nonzero interference term, which will be zero in a classical environment (i.e. when the operators ρ_C and $\overline{\rho_C}$ coincide). In the case of a pure prior state, $\delta = 0$ for a first agent, however, for the second agent, the paper shows calculation of δ not to be necessarily zero. Khrennikov [42] says (p. 99): "Such single agent interference can be interpreted as interference between her information representation, a prior state, and common knowledge about posterior probabilities. Thus agreement on disagreement is possible because the agent having the common prior state ρ and having common knowledge about posterior probabilities can interfere in different ways with these commonalities."

References

[1] Khrennikov, A. (2023). Open systems, quantum probability, and logic for quantum-like modelling in biology, cognition and decision-making. *Entropy*, 25, 886, 1–24.

[2] Plotnitsky, A. and Haven, E. (2023). *The Quantum-Like Revolution: A Festschrift for Andrei Khrennikov*. Cham: Springer Nature

[3] Tversky, A. and Kahneman, D. (1983). Extensional versus intuitive reasoning: the conjunction fallacy in probability judgment. *Psychological Review*, 90, 4. https://doi.org/10.1037/0033-295X.90.4.293

[4] Pothos, E. and Busemeyer, J. (2022). Quantum cognition. *Annual Review of Psychology*, 73, 749–778.
[5] Busemeyer, J. R., Pothos, E. M., Franco, R. and Trueblood, J. S. (2011). A quantum theoretical explanation for probability judgment errors. *Psychological Review*, 118(2), 193–218.
[6] Tversky, A., Shafir, E. (1992). The disjunction effect in choice under uncertainty. *Psychological Science*, 3, 305–309.
[7] Savage, L. J. (1954). *The Foundations of Statistics*. Hoboken, NJ: John Wiley & Sons.
[8] Busemeyer, J., Wang, Z. (2007). Quantum information processing: explanation for interactions between inferences and decisions. In *Quantum Interaction* (Stanford University). Papers from the AAAI Spring Symposium. Technical Report SS-07-08. Bruza P. D., Lawless, W., van Rijsbergen, K. and Sofge, D. (eds.); pp. 91–97.
[9] Haven, E., Khrennikov, A., Robinson, T. (2017). *Quantum Methods in Social Science: A First Course*. Singapore: World Scientific Publishing.
[10] Barkan, R. and Busemeyer, J. R. (2003). Modeling dynamic inconsistency with a changing reference point. *Journal of Behavioral Decision Making*, 16, 235–255.
[11] Haven, E. and Khrennikov, A. (2017). *The Palgrave Handbook of Quantum Models in Social Science*. London: Palgrave-McMillan Publishers.
[12] Busemeyer, J. R., Wang, Z. and Townsend, J. T. (2006). Quantum dynamics of human decision-making. *Journal of Mathematical Psychology*, 50(3), 220–241.
[13] Busemeyer, J. R., Bruza, P. D. (2012). *Quantum models of cognition and decision*. Cambridge, UK: Cambridge University Press.
[14] Varian, H. R. (1992). *Microeconomic Analysis*. New York: W.W. Norton & Company.
[15] von Neumann, J., Morgenstern, O. (1994). *Theory of Games and Economic Behavior*. Princeton, NJ: Princeton University Press.
[16] Allais, M. (1953). Le comportement de l'homme rationnel devant le risque. Critique des postulats et axiomes de l'école Américaine. *Econometrica* 21, 503–546.
[17] Kreps, D. (1988). *Notes on the Theory of Choice*. Underground Classics in Economics. Boulder, Colorado: Westview Press.
[18] Karni, Ed (2005). *Savage's subjective expected utility model*. Baltimore, MD: Johns Hopkins University: www.econ2.jhu.edu/people/Karni/savageseu.pdf
[19] Ellsberg, D. (1961). Risk, ambiguity and the Savage axioms. *Quarterly Journal of Economics*, 75: 643–659.
[20] Anscombe, F. J., Aumann, R. J. (1963). A definition of subjective probability. *Annals of Mathematical Statistics*, 34, 199–205.
[21] Knight, F. H. (1921). *Risk, Uncertainty and Profit*. Boston, MA: Houghton Mifflin.
[22] Aerts, D., Haven, E. and Sozzo, S. (2018). A proposal to extend expected utility in a quantum probabilistic framework. *Economic Theory*, 65, 1079–1109.
[23] Gilboa, I., Schmeidler, D. (1989). Maxmin expected utility with a non-unique prior. *Journal of Mathematical Economics*, 18, 141–153.
[24] Klibanoff, P., Marinacci, M. and Mukerji, S. (2005). A smooth model of decision making under ambiguity. *Econometrica*, 73, 1849–1892.
[25] Etner, J., Jaleva, M. and Tallon, J. M. (2012). Decision theory under ambiguity. *Journal of Economic Surveys* 26, 234–270.
[26] Gilboa, I., Marinacci, M. (2013). Ambiguity and the Bayesian paradigm. In Acemoglu, D., Arellano, M. and Dekel, E. (eds.) *Advances in Economics and Econometrics: Theory and Applications*. New York, NY: Cambridge University Press.
[27] Machina, M. J., Siniscalchi, M.: Ambiguity and ambiguity aversion. In: Machina, M. J., Viscusi, K. (eds.), *Handbook of the Economics of Risk and Uncertainty*, pp. 729–807. New York, NY: Elsevier.

[28] Ellsberg, D., Machina, M., Ritzberger, K. and Yannelis, N. (Eds.) (2011). Symposium on the 50th anniversary of the Ellsberg Paradox. *Economic Theory* 48, 219–548.
[29] Machina, M. J. (2005). Expected utility/subjective probability analysis without the sure-thing principle or probabilistic sophistication. *Economic Theory* 26, 1–62.
[30] Chateauneuf, A.; Eichberger, J.; Grant, S. (2007). Choice under uncertainty with the best and worst in mind: neo-additive capacities. *Journal of Economic Theory*, 137, 538–567.
[31] Denuit, M.; Dhaene, J. Goovaerts, M., Kaas, R., Laeven, R. (2006). Risk measurement with equivalent utility principles. *Statistics and Decisions*, 24, 1–25.
[32] Kelsey, D. (1994). Maxmin expected utility and weight of evidence. *Oxford Economic Papers*, 46(3), 425–444.
[33] Machina, M. J. (2009). Risk, ambiguity, and the rank-dependence axioms. *American Economic Review*, 99, 385–392.
[34] Aerts, D., Sozzo, S. (2014). Quantum entanglement in concept combinations. *International Journal of Theoretical Physics*, 53, 3587–3603.
[35] Wang, Z.; Solloway, T.; Shiffrin, R. M. and Busemeyer, J. (2014). Context effects produced by question orders reveal quantum nature of human judgments. *PNAS*, 111 (26) , 9431–9436.
[36] Yearsley, J. (2021). *The Quantum Question Equality*. Nashville, TN: Vanderbilt University.
[37] Pothos, E., Busemeyer, J. (2022). Quantum cognition. *Annual Review of Psychology*, 73, 749–778.
[38] Ozawa, M., Khrennikov, A. (2021). Modelling combination of question order effect, response replicability effect, and QQ-equality with quantum instruments. *Journal of Mathematical Psychology*, 100, 102491.
[39] Aumann, R.(1976). Agreeing to disagree. *Annals of Statistics*, 4, 1236–1239.
[40] Hild, M., Jeffrey, R. and Risse, M. (1996). A generalization of Aumann's agreement theorem. Princeton, NJ: Princeton University. www.princeton.edu/~bayesway/Dick .tex.pdf.
[41] Morris, S. (1995). The common prior assumption in economic theory. *Economics and Philosophy*, 11, 227–253.
[42] Khrennikov, A. (2015). Quantum version of Aumann's approach to common knowledge: Sufficient conditions of impossibility to agree on disagree. *Journal of Mathematical Economics*, 60, 89–104.

6

Bohmian Mechanics and Finance

6.1 Probability Embedded (or Not) in the Social Science Model

The quantum-like community has emphasized the processing of information as one of the mainstays of quantum physics (and also of quantum-like applications). There is one qualifier, though: this is not information processing in the sense of quantum computers or quantum cryptography. In fact, the goal consists in applying the general information processing framework to areas like economics and other areas of the social sciences, biology, and the humanities.

Bohmian mechanics, by virtue of its notion of quantum potential and its relationship with a measure of information, Fisher information, plays a central role in the information modelling objective we set in social science. Moreover, if Bohmian mechanics is adoptable in social science it will also auto-classify itself in a model where probability is inherently part of the model. We make this assertion, as we think that many models in social science do just the opposite by instead having the model inform probability.

6.1.1 Examples of models informing probability

Let us give some examples of such models where probability is not an inherent part of the model per se, but rather where the model steers or informs the probability.

1. Rational expectations: we have discussed this well-known approach in economics in Chapter 1. Here the economy is represented as being one individual. Given the obvious uncertainty of future prices, within rational expectations, decision makers are assumed to have identical distributions on those future prices. This is thus an example where the model informs the probability: (i) assume identical distributions and therefore (ii) the distribution shall be...
2. In Kirman's ant [1] [2] model the Ehrenfest urn model is used when ants do not interact with each other to go to a food source. However, when the ants

do interact to varying degrees, his model is the source for different probability outcomes: a martingale for instance. The model informs the probability.

3. The financial option pricing model renders the price of call option to be the difference between two cumulative distribution functions. Again the model informs the probability. Inherent in the setup of option pricing models is the construction of risk neutral probabilities, which are nonobserved. The model informs the probability.

6.1.2 Probability as an Inherent Part of the Model

In past chapters we have talked about the quantum mechanical wave function. This is not a classical wave: it has no energy and lives in a space which is not the three-dimensional real space. As Heisenberg once said - the probability wave stands "in the middle between the idea of an event and the actual event." But what is this wave function then within a social science context? We can say the following:

1. The wave function unifies all contextual measurements.
2. There is contextual interaction between what is measured and the measuring apparatus.
3. A measurement outcome is actualized among a set of possible outcomes, as a consequence of this contextual interaction.
4. This contextual interaction is noncontrollable (and generates the Heisenberg uncertainty principle).

Out of the preceding factual observations, we make the premise that human judgments and decisions are constructive processes in which the interaction between the object of the decision and the decision maker is of a cognitive nature (rather than physical in nature).[1] and this brings us back to (possibly) the idea that quantum probability expresses a deeper level of uncertainty. We discussed the possible characterization of uncertainty in Chapter 5. Furthermore, in that same Chapter 5 we discussed the Ellsberg paradox, and here the wave function approach (quantum-like approach) can accommodate observed percentages on two-stage sequential game models by using the probability interference term. Probability is surely inherent in the model.

George Soros in [4] argues that in his theory of reflexivity, the economic agent's thinking has two functions:

1. "to understand reality; that is the cognitive function"
2. "to make an impact on the situation...(the) manipulative function"

[1] See special issue in *Journal of Mathematical Economics* [3]

If both functions, in the words of Soros, "operate simultaneously (then) they interfere with each other." It is this interference that Soros calls 'reflexivity.' The key here resides in the fact that that if both functions interfere, they produce uncertainty in both the participant's understanding and the actual course of events. The interference works as a feedback loop: positive feedback reinforces both the prevailing trend and the prevailing bias and leads to a mispricing of financial assets, whilst the negative feedback corrects the bias.

The wave function can come into this model in the following way. When bringing reflexivity into the realm of wave function construction, we could identify the existence of a negative (or positive) feedback with the width of an amplitude function. The width of the resultant wave function, we will propose, can be used as an indicator of the degree of erroneous information. Under negative feedback, the amplitude function on the wave number would be very wide and, via Fourier integration, this would create a very tight wave function. If the wave function is dependent on the position parameter, which is price, then the density function derived from that wave function will be tight and the probability of mispricing would be small. However, under positive feedback, the amplitude function on the wave number would be very tight and, via Fourier integration, this would create a very wide wave function. If again, this wave function is dependent on price, the probability of mispricing would indeed now be high.

Information here is central in this approach and probability is inherent in the model. In the following sections of this chapter we will delve into how Bohmian mechanics avails itself of a degree of usefulness which really should not be underestimated. The wave function is still linked to probability but now also guides the particle. Probability here in this approach is used to reflect the ignorance, due to the confluence of factors, of where precisely the electron will be.

Before we embark on a deeper discussion (in this chapter and the two next chapters), we want to give a few cautionary stopsigns (which we do not believe to be a problem, however):

1. Conservation is an issue.
2. Time reversibility is an issue.
3. Hermiticity is an issue.

6.2 Bohmian Mechanics and Finance: The Groundwork

In this section we will highlight details on the Bohmian mechanics approach in finance. We explicitly use the word *groundwork*, as we believe that for the purposes of applying Bohmian mechanics in finance, the approach by Hawkins and Frieden, detailed in the following subsection, lays out an elegant and (from an economic point of view) a quite palatable approach to *why* it is reasonable to consider this type of mechanics in economics and finance. Our argument does not refer to the fact that Bohmian mechanics gives a quasi-classical interpretation to quantum mechanics. Rather, it is the use of the Lagrangian optimization approach, so common in economics, which immediately brings a feeling of familiarity to any economist contemplating the form of Equation (6.2). The centrality of the information measure, which is 'trigger started' in Equation (6.1), is another key ingredient which makes this approach enticing to economic theory. We want to stress that we have here an information-based approach, right from the outset. The approach is squarely physics based but also really adds to our understanding of modelling economics processes with an information-centric model.

This *groundwork* section not only discusses the Hawkins and Frieden approach, it also documents the very first published papers which applied, using financial data, the Bohmian approach. The very first idea of using Bohmian mechanics in social science was proposed by Andrei Khrennikov. The first applied paper was by Tahmasebi et al. Those papers are also briefly discussed in this section.

6.2.1 The Hawkins and Frieden Approach

An important contribution to the area of financial economics has been made by the work of Raymond Hawkins and Roy Frieden. Both authors emphasize the rationale for the existence of quantization in finance. Their point of view can be summarized in the following quote: [5] "A key question, however, is whether quantization in economics is simply a convenient analytical consequence of the well-known stochastic representation of financial economics or if the link between quantum mechanics and economics has an alternative origin. Our work suggests a more fundamental information-theoretic basis for quantization in economics." Their comment narrowly allies to what we said in the introduction of the previous chapter when we discussed the quantum-information revolution.

Their comment is an important one, especially if we position it relative to the historical development of physics. A lesser known fact in the development of quantum physics may well be the work of Einstein and others on how quantum mechanics can be seen as, as Andrei Khrennikov [6] calls it (p. 4) "a more general quantum theory with nonlinear dynamic equations." Andrei Khrennikov says, "The most difficult problem of such nonlinear quantum theories is the formulation of a nonlinear analog of the superposition principle." See also [7]. If one were to think of such a theory as successful, what would quantization then look like?

A central premise that the authors start from is that economic dynamics follow from information flow perturbations. The extremization of Fisher information leads to an equation of motion which is isomorphic to the Schrödinger equation. This is a very strong result. First of all, Fisher information has not only relevance in physics, more particularly in Bohmian mechanics but it also has a key role to play, albeit under a derived notion well known in economics under the name of 'Cramer–Rao bound'. Fisher information in Hawkins and Frieden is defined as:

$$I = \int \frac{1}{P}\left(\frac{\partial P}{\partial x}\right)^2 dxdt = 4\int |\psi'(x)|^2 dxdt. \tag{6.1}$$

As is well known with the Fisher information measure, information is high when volatility around the variable in question (f.i. price) is low (and vice versa).

Hawkins and Frieden [5], p.23 formulate the Lagrangian:

$$\mathcal{L} = \int \frac{1}{P}\left(\frac{\partial P}{\partial x}\right)^2 dx + \lambda_0\left[1 - \int Pdx\right] + \sum_{m=1}^{M} \lambda_m \left[d_m - \int f_m P dx\right], \tag{6.2}$$

and they also consider a key equation:

$$\frac{d^2\psi(x)}{dx^2} = -\frac{1}{4}\left[\lambda_0 + \sum_{m=1}^{M} \lambda_m f_m(x)\right]\psi(x), \tag{6.3}$$

where the Lagrange multiplier λ_0 corresponds to the normalization condition for the probability density ($\int p(x)dx = 1$).

The preceding equation is of the same format as the (time-independent) Schrödinger PDE:

$$-\frac{\hbar^2}{2m}\frac{d^2\psi(x)}{dx^2} = \left[E + V(x)\right]\psi(x). \tag{6.4}$$

What is important to remark is that if, for instance, $f_m(x) = x$ and x is the price of an asset, then one writes an expectation of that price in the preceding Schrödinger-like equation. Hawkins and Frieden [5] also remark, and this is important, that, for instance, for $f_1(x) = x$ with the Lagrange multiplier λ_1 forms the potential function. We note that, for instance, $f_m(x)$ could take on another financial meaning such as the intrinsic value of an option contract. We also mention that Hawkins and Frieden [5] (pp. 30–33) continue to discuss also a financial analogue of the time-dependent Schrödinger PDE. Hawkins and Frieden argue that quantization is steeped in an information-theoretic basis.

6.2.2 The First papers using Bohmian Mechanics in Finance

Andrei Khrennikov was the first author to remark how Bohmian economics could be used in social science applications (see Khrennikov [8]). Moreover, the first paper which documented the (financially relevant) stochastic aspects of using Bohmian mechanics in finance, was by Choustova, [9]. As we remarked in the introduction to this section, the path availability which is an advantage of using Bohmian mechanics, is unfortunately proving to be a downside from a stochastic point of view. One of the issues with the use of Bohmian mechanics in finance is that the trajectories which are generated via the extended second law of Newton do not exhibit nonzero quadratic variation. In mathematical finance, price paths do exactly exhibit nonzero quadratic variation. Trajectories in Bohmian mechanics are smooth. We refer the interested reader to the paper by O. Choustova, [9], where this issue is addressed in great detail.

The first paper which applies Bohmian mechanics to finance, per se, is by Tahmasebi et al., [10]. The authors compare the quantum potential for the Standard and Poor's stock index with the quantum potential of a corresponding Gaussian white noise with the same variance; where the white noise concept corresponds to a community which is not biased by the prices prior to its date. Mathematically speaking, the function R^2 for a white noise which is a Gaussian function represented by:

$$\exp(-(q - q_0)^2)/2\sigma^2 \tag{6.5}$$

when substituted in the quantum potential expression, gives the quantum potential of the white noise,

$$U = \left[\frac{(q - q_0)^2}{\sigma^4} - \frac{1}{\sigma^2}\right], \tag{6.6}$$

where q_0 represents the average variations of the real market price returns under comparison, and σ^2 is the variance. Note that q_0 and σ take values from their corresponding real markets. The authors find some very interesting results such as the wall distance of the potential function relates to the amplitude variation of the price returns.

Subsequent to the paper by Tahmasebi et al., we note the work of Khaksar et al., [11], where the so-called Q-risk is defined as being proportional to the distance between two walls of the square well potential. This type of risk is heuristically defined, within the setting of portfolio management, as the wall width of the quantum potential on returns.

The paper argues for the source of information relative to the quantum versus the real potential. We argue that the information sitting in the higher moments of the probability density function can be captured by the quantum potential. Hence, we could argue that with Gaussian probability density functions, the quantum potential will not contribute more information relative to the real potential. We remark that the Gaussian probability density function is defined relative to the real potential. Finally, the quantum and real potentials can in fact only be compared in terms of their shapes since there are scaling issues which would need accounting for.

[2] R stands for the amplitude function of the wave function.

Hosseini et al. in [12] carried the work further to mostly where the current state of affairs is now with using Bohmian mechanics in finance. In this paper, the quantum potential for Gaussian and non-Gaussian densities is constructed and the impact of rare events on those quantum potentials is studied. The quantum potential does clearly show, intuitively, how extreme events do become much more apparent when the level of non-Gaussian scaling is augmented. Being able to characterize this is an important achievement in the literature on applying Bohmian mechanics to finance, especially if it can be built into the construction of price paths.

6.3 Bohmian Mechanics and Finance: Financial Information

Chen and Haven in [13] first consider finding the extracted density function for daily returns on several highly traded commodities. The correlations between both forces, using the real and quantum potential, do show a weak level, within a range of the measured returns on the commodities. This may indicate that there are different types of information in operation, all sourced from the same information type: namely public information. It is also clear from that paper, that correlations between forces out of the range of feasible returns (i.e. at the walls of the potentials) are high. Here information types now coincide?

The preceding paper also finds quite consistently, for the period 1995–2017, that the wall distance on the quantum potential on major stock indices is larger for stock indices of developing markets as opposed to those of developed markets. This is a very intuitive result. Developing markets should exhibit higher variability than developed markets. This wall distance is also confirmed for well established indices like the NASDAQ versus the Dow Jones, respectively wider and tighter, as we would expect.

The Kernel density estimation (KDE) method is used for that purpose. If the probability density function is time-independent, then we can extract the real potential from:

$$\frac{1}{2}Q\frac{\partial f(q)}{\partial q} = Kf(q). \tag{6.7}$$

This ODE is solvable by the following:

$$f(q) = N \exp\left(-\frac{2V(q)}{Q}\right). \tag{6.8}$$

The real potential is then

$$V(q) = -\int K(q)dq, \tag{6.9}$$

where N is determined by $\int_{-\infty}^{+\infty} f(q)dq = 1$.

Furthermore, the quantum potentials for return time series of commodities in different time frequencies is then also derived. Using the usual derivation of the quantum potential where the polar form of the wave function is inserted into the Schrödinger PDE, one of the key equations which is obtained is

$$\frac{\partial S}{\partial t} + \frac{1}{2m}\left(\frac{\partial S}{\partial q}\right)^2 + \left(V - \frac{h^2}{2mR}\frac{\partial^2 R}{\partial q^2}\right) = 0. \tag{6.10}$$

The probability density function is

$$R^2 = |\psi|^2 \tag{6.11}$$

and the last term in the preceding equation is the quantum potential. The paper does find two useful results:

1. Real and quantum potential are different objects: a potential well is found for commodity returns; whilst square wells are found for the quantum potential.
2. The force when calculated from the negative gradients on either the real potential or the quantum potential do show coincidence at the level of the walls of both potentials. However, in between the walls, forces do behave differently. A weaker negative reaction trend in the case of quantum potential (relative to real potential) seems to occur. The source of information is the same for both potentials: it is the same public information. The in-range values (between the coinciding forces occurring at the walls of the potentials) are thus different. They capture a different dimension of that common public information. We refer the reader to the Rosenow model (in Chapter 4) for a possibly more structured answer to this problem.

6.4 Quantum Pilot-Wave Theory and a Macroscopic Fluid System

A few years ago an important paper by John Bush, [14], considered the so-called 'walking droplets' which he showed were not confined to quantum mechanics. The experiments which were performed on a fluid bath could, we believe, imply promise that the quantum formalism may lend characteristics which are (maybe) findable on a macroscopic scale. In a double-slit experiment, the walker droplet passes through one slit or the other. The guiding waves pass through both slits: i.e. the walker droplet 'feels' the second slit by virtue of the pilot wave (see Bush in [14]). The motion of a droplet is driven by an interaction with a superposition of waves emitted by the points the droplet has visited previously.

Brady and Anderson in [15] show that the droplets follow an analogue of the Schrödinger equation (with a constant of motion which is not h). Theo Nieuwenhuizen in [16] mentions that this 'quantum-like' behavior, as found in the lab, "proves the possibility that true quantum behavior originates from classical stochastic forces."

Clearly, this can be an exciting avenue to pursue further within the setting of social science (in finance more specifically). However, is the observed behaviour in the fluid bath really amenable to the use of some quantum formalism? In the paper, the walker system (obtained via millimetre-sized drops on a carefully attuned vibrating fluid bath) has high path memory. Even in reference to tunnelling, the paper [14] mentions that (p. 274): "the quantum-like behavior (Gamow in [17]) results from the pilot-wave dynamics, and the unpredictability is rooted in the complex interaction between the droplet and its spatially extended guiding wave."

The author in [14] asks a pertinent question toward the end of the paper (p. 283) concerning "what relation this hydrodynamic system might have to existing realist models of quantum dynamics." At this point in the book, it may be appropriate to make a distinction between at least three difference stances one could take toward 'extending' one's understanding and/or use of quantum mechanics. Here they are:

- (1) is there (a) (or are there) real dynamic(s) underlying quantum theory?
- (2) is there a 'subquantum' theory?
- (3) is there a 'quantum-like' theory?

Starting in reverse order, item (3) is to a large extent the main theme of the book. As we have already discussed in the beginning of Chapter 5, in essence one can propose to use the formalism of quantum mechanics outside of quantum mechanics, without making a claim that an aspect of the macroscopic world is 'inherently' quantum mechanical. As we have tried to make clear earlier, there are, for instance noncommutativity occurrences in the decision-making field which open the door to the idea of non-commutativity relevant to the intersection of events, and the ensuing use of quantum probability. The orientation exemplified in item (2), we believe, is probably best highlighted by the reference that Theo Nieuwenhuizen in [16] makes to the work of De Raedt and Michielsen. Nieuwenhuizen says in [16], p. 6: "De Raedt and Michielsen [19] have designed learning algorithms for event-by-event simulation, which reproduce the quantum predictions for many photon and neutron experiments without any quantum input." We note thus that the 'subquantum' approach, whilst it clearly focuses on quantum mechanics per se, is not concerned whether real quantum dynamics are underlying the statistically based quantum theory. The question of real quantum dynamics is then posed in item (1), and it rejoins the question Bush asked in [14] (and which we repeated earlier), namely how the hydrodynamic system can have links to quantum dynamics. The interested reader is referred to especially Section 4.1. (dealing with the Madelung transformation in [18]) and the discussion in the following section, where the (not so) closeness of the walker system with Bohmian mechanics is expounded upon.

The paper finishes with an excellent discussion on how really close this hydrodynamic system is to quantum mechanics. As the author mentions (p. 287), although intrusive measurement can be imposed in this setting, he also does caution on whether a hydrodynamic analogy of entanglement may exist (or not exist).

We like stressing, however, that we believe the hydrodynamic interpretation to be a strong starting point for information modelling in economics and finance. Further on in the book, in Section 7.7, we highlight some possible applications when we discuss a memory model.

References

[1] Kirman, A. (1993). Ants, rationality and recruitment. *The Quarterly Journal of Economics*, 108(1), 137–156.

[2] Kirman A. (2017). The economy as a complex system. In Aruka, Y. and Kirman, A, *Economic Foundations for Social Complexity Science.* Singapore: Springer.

[3] Haven, E., Khrennikov, A., Ma, C. and Sozzo, S. (2018). Special issue on quantum probability in Economics: Introduction to quantum probability theory and its economics applications. *Journal of Mathematical Economics*, 78, 127–130.

[4] Soros, G. (1987). The alchemy of finance. Reading the mind of the market. New York: John Wiley & Sons.

[5] Hawkins, R., and Frieden, R. (2017). Quantization in financial economics: an information-theoretic approach. In Haven, E., and Khrennikov, A. *The Palgrave Handbook of Quantum Models in Social Science*. London: Palgrave Macmillan.

[6] Khrennikov, A. (2023). Open systems, quantum probability, and logic for quantum-like modelling in biology, cognition and decision-making. *Entropy*, 25, 886, 1–24.

[7] Bialynicki-Birula, I. and Mycielsky, J. (1976). Nonlinear wave mechanics. *Annals of Physics*, 100, 62–93.

[8] Khrennikov, A. Y. (1999). Classical and quantum mechanics on information spaces with applications to cognitive, psychological, social, and anomalous phenomena. *Foundations of Physics*, 29(7), 1065–1098.

[9] Choustova, O. (2008). Quantum model for the price dynamics: the problem of smoothness of trajectories. *Journal of Mathematical Analysis and Applications*, 346, 296–304.

[10] Tahmasebi, F., Meskinimood, S., Namaki, A., Farahani, S.V., Jalalzadeh, S. and Jafari, G.R. (2015). Financial market images: a practical approach owing to the secret quantum potential. *Europhysics Letters*, 109(3), 30001.

[11] Khaksar, H., Haven, E., Nasiri, S. and Jafari, G. (2021). Using the quantum potential in elementary portfolio management: some initial ideas. *Entropy*, 23, 180.

[12] Hosseini, R., Tajik, S., Koohi Lai, Z., Jamali, T., Haven, E. and Jafari, R. (2023). Quantum Bohmian-inspired potential to model non-Gaussian time series and its application in financial markets. *Entropy*, 25, 1061.

[13] Chen, S. and Haven, E. (2017). Using empirical data to estimate potential functions in commodity markets: some initial results. *International Journal of Theoretical Physics*, 56, 4092–4104.

[14] Bush, J. W. M. (2015). Pilot wave hydrodynamics. *Annual Review of Fluid Mechanics*, 47, 269–292.

[15] Brady, R. and Anderson, R. (2014). Why bouncing droplets are a pretty good model of quantum mechanics. *Arxiv: 1401.4356v1.*
[16] Nieuwenhuizen, Th. (2004). A subquantum arrow of time. *arxiv: 1409.3131v1.*
[17] Gamow, G. (1928). The theory of nuclear disintegration. *Nature*, 122:805–7.
[18] Madelung, E. (1926). Quantentheorie in Hydrodynamischen form. *Zeitschrift fur Physik*, 40:322–26.
[19] De Raedt, H., Zhao, S., Yuan, S., Yin, F., Michielsen, K. and Miyashita, S. (2010). Event-by-event simulation of quantum phenomena. *Physica E*, 42(3), 298-302

7
Other Quantum-like Models in Finance

The noncompatibility of observables refers back to Bohr's complementarity principle. It is a key idea in quantum mechanics but it can also translate into other areas outside of physics. The nonjoint observation of price and the time change of price was already remarked by Belal Baaquie [1]. We have made the argument that precisely this incompatibility leads to understanding why arbitrage[1] occurs. Baaquie also has argued that the quantum Hamiltonian out of Black–Scholes option pricing is non-Hermitian[2] and this observation is narrowly related to the martingale condition which needs to be satisfied. In this chapter, besides non-Hermiticity, we also consider the use of quantum harmonic oscillators and quantum probability in finance and give a brief appreciation of research on using quantum field theory in finance. We also consider in this chapter other applications, notably in seeing how investment choices can be mapped onto a sequential gambling problem which we discussed earlier in relation to the violation of the Ellsberg paradox. We also discuss an application on data retrieval as it relates to incomplete financial information. We close the chapter with some thoughts on both the expansion and limitations of mathematical modelling.

7.1 Quantum Harmonic Oscillators and Quantum Probability

In this section, we begin with one approach which considers demand and supply from a quantum harmonic oscillator point of view. In the next section, we then continue discussing option pricing via the use of quantum probability. None of those approaches consider a particular acceptance of an interpretation to quantum mechanics. In that regard they are quite different from the way we approach, in this book, the use of quantum mechanics to social science. Thus, the emphasis here is still on the formalism (for instance: quantum probability or quantum oscillators).

[1] Arbitrage can roughly be defined as making a riskless profit.

[2] Hermiticity is a key property in quantum mechanics.

David Orrell [2] introduced the idea of using quantum harmonic oscillators in finance. A central theme in his work revolves around the idea that threshold effects are quite common in social science phenomena. As David Orrell in [2] says (p. 325): "These threshold phenomena can be understood using methods based on quantum probability... which leads inevitably to the notion of a quantum of action or of energy." And as the author continues, this quantum of action relates to the Planck constant for which no equivalent in quantum-like applications has yet been pinpointed. Central to his paper is the idea of a quantum of value. The entropic force is defined by Orrell in [2] as

$$F(x) = \gamma P'(x)/P(x), \tag{7.1}$$

where γ is a scaling parameter. We note that the author defines $P(x)$ as a so-called propensity function which has a peculiar interpretation: [2], p. 326: "the probability distribution $P(x)$ which specifies the probability of the person buying or selling at a particular price x, where x is usually chosen to be a logarithmic variable. The resistance to changing to some nearby price $x + \Delta x$ will depend on the change in propensity conditional on (or relative to) the current propensity." And as the author further clarifies (p. 326): "In economics, the force represents the mental desire for a buyer or seller to adjust the price to their own preferred level. The work done in a change of log price from x_1 to x_2 is

$$\Delta E = \int_{x_1}^{x_2} F(x)dx = \gamma \log\left(\frac{P(x_2)}{P(x_1)}\right) \tag{7.2}$$

which depends only on the relative initial and final propensities (so it is not necessary to normalize the propensity function)." As Orrell remarks, there is an immediate connection to supply and demand and it is indeed a relation which even intuition tells us must exist. But as one can remark: what is so quantum (or quantum-like) here? Indeed as Orrell argues (p. 326), "the price does not behave in a classical fashion: it is indeterminate until observed through a transaction, so transactions act as a process of measurement (see Sarkissian in [3])."

Orrell [2] (p. 327) thus then gets to a quantized version of a supply/demand model: (see Orrell in [4]) which is a quantum harmonic oscillator with a mass:

$$m = \frac{\hbar}{2\omega\sigma^2} \tag{7.3}$$

and force constant:

$$k = m\omega^2 = \frac{\hbar\omega}{2\sigma^2}. \tag{7.4}$$

7.2 Quantum Option Pricing with the Aid of Quantum Probability

In the work by Sergio Focardi and Frank Fabozzi and Davide Mazza [5] the authors write that (p. 2): "in finance theory there are strong motivations that might suggest adopting quantum probability:

- Quantization of variables and economic states (i.e. observations can assume only discrete values)
- Self-reflexivity of markets so that observations influence financial markets
- Incompatibility of pairs of observations
- Identification of observations with trades and money transfer."

In a similar vein to the work of David Orrell in the preceding section, the authors specify essential differences between classical and quantum probability, one of them being that (p. 3) "Uncertainty is not due to lack of information; it is native and cannot be eliminated."

As the authors remark, in physics the rationale for using quantum probability is well established. As the authors say (p. 11): "In physics, quantum probability is strongly recommended by empirical observations, in particular experiments of diffraction and polarization. The theoretical work of John Stewart Bell at CERN (Bell's inequalities) and the experiments performed by Alain Aspect[3] at the Ecole Supérieure d'Optique in Orsay show that we need a new probability theory unless we are ready to accept transmission of signals faster than light. Quantum probability is needed because of empirical results in physical experiments."

However, as the same authors remark, there is in finance (p. 10) "no crucial experiment, or crucial empirical evidence in favour of quantum probability. Quantum probability is recommended primarily by indirect evidence." As we have also remarked earlier in the book (and the authors of this paper also argue for it), the self-reflexivity of markets (i.e. the Soros approach), amongst others, can be invoked to argue for the use of quantum probability in finance.

In option pricing, two key approaches which always need mentioning are the Segal and Segal approach and the ensuing Accardi and Boukas stance. As Focardi et al. remark in [5], p. 17, "Segal and Segal [6] claim that a quantum variant of this theory serves to facilitate the modeling of phenomena not fully explained by it, such as (1) short-term volatility, (2) extreme dis-continuities and serial correlation, and (3) large-tailed distributions." The authors also do correctly mention that the perennial issue of simultaneous measurement of prices and their instantaneous derivatives is at the core of the quantum approach. This has also been remarked by other authors such as Belal Baaquie [1]. Focardi et al. in [5] describe the Accardi

[3] The authors of the book add that Professor Aspect, together with Professors Zeilinger and Clauser, were the 2022 winners of the Nobel Prize in Physics.

and Boukas approach [7] where the authors replace (p. 17) "Brownian motion by two annihilation and creation operators in the Fock space."

We like to emphasize how Focardi et al. argue for the benefits of adopting the quantum framework in finance. The authors say (p. 18): "The benefits of adopting a quantum framework are twofold. First, it provides better handling of uncertainty, especially a better handling of extreme events. Second, a quantum framework is better equipped than the classical framework in providing an early warning of large or even catastrophic events."

7.3 Quantum Probability Applied to Investment Choice

In the work by Haven and Khrennikova an experiment (with student participants) [8] is introduced to attempt to establish a quantum-probability-based framework which can formalize investment preferences.

So-called portfolio games were given an experimental flavour in that (under)graduate students enrolled in a business school participated in an experiment to yield investment preference elicitation. This elicitation is essentially based on the two-stage gambling experiment discussed earlier in the book. This is the gambling experiment which consistently shows the occurrence of the violation of the sure-thing principle. As is explained in the paper, the risk perception, though maybe different in that the setting here is now based on an emulation of real-world financial decisions, of course, the two-stage gamble is simply based on a gambling setting.

The experimental design is described as follows: ([8], p. 190): "Both in the initial 'Pilot experiment' and the 'Main experiment', the investment task was presented to the participants as a portfolio game with a reinvestment opportunity for the second investment round." ... The experiment lines participants needed to digest before they started playing were as follows[8], pp. 190–191: "Imagine that you are an investor on the financial market. You have borrowed £1000 to invest in a portfolio and will have to return it at the end of the portfolio game. Please neglect interest rates on the £1000. You can own the positive return the portfolio might make. Equally, you could obtain a negative return on your portfolio investment. Assume the time is now 9:00 am. Consider two future times: 10:00 am and 11:00 am. The portfolio is predicted to have a strictly 50 percent chance of obtaining a + 20 percent profit and 50 percent chance to generate a loss of −10 percent at 10:00 am. Equally, the portfolio has a 50 percent chance of obtaining a + 20 percent profit and 50 percent chance of having a loss of −10 percent at 11:00 am. Consequently, you can either gain £200 or lose £100 at 10:00 am and 11:00 am. You can only acquire information about the realized portfolio return from a portfolio manager. This means you cannot obtain information about the portfolio's price change from any other source (internet, newspapers, etc.). At the same time, the portfolio manager has a purely informative role and cannot influence the price of the portfolio".

As is the case with the two-stage gamble problem, information is either released that one won or lost – or no information is released. There is a variant in the experiment, where participants are served with the full scenario, upfront: that is, they are asked what they would do – if they had information they won, or information they lost, or had no information at all. As was explicitly remarked in the paper, the result of the experiments show the existence of a disjunction effect but there were only minor cases of preference reversals. In the case where all participants are faced with all the information, upfront, the investment preference context yielded the same results as reported in Tversky and Shafir [9] and Croson [10] (where no finance context is used at all).

The paper also ventured into formalizing a quantum-like description of a representative agent's preference formation. For that purpose, a set of acts $f_1, f_2 \in F$ corresponding respectively to 'play' and to 'quit' were defined, and likewise, the states of the world $s_1, s_2 \in S$ were then mapped into the 'Won' or 'Lost' setting in the portfolio experiment. As was remarked in [8] , p. 193, "The marginal probability of a person in a NYK (no information) setting to choose some option f_j should be equal to the sum of disjunctions of f_j conditioned upon the events s_1, s_2 (in our setting, a loss, or a gain in a previous period of the portfolio game)." The results from the experiment show indeed that the sure-thing principle is not followed all the time.

7.4 Quantum Probability and Process Mining

In another finance application, the paper by Moreira et al., [11], considers the use of nonclassical probability to retrieve missing financial loan data. This paper looks at a real-world problem where there is missing data in financial bank loan applications. The starting point of the paper revolves around the idea of so-called 'process mining', which in the paper (p. 2) is defined as per [12] : "a technique that enables the automatic analysis of business processes based on event logs. Instead of designing a workflow, process mining consists in gathering the information of the tasks that take place during the workflow process and storing that data in structured formats called event logs." Relative to process mining, the authors make the argument that Bayesian networks can manage uncertainty as opposed to Markov chains, where events need all occurring. Their work really targets the using of quantum-like Bayesian networks within the missing data loan application process. They define such Bayesian networks as: (pp. 16–17) "an acyclic directed graph in which each node represents a random variable. Each edge represents a direct influence from the source node to the target node and uses probability amplitudes, which will be responsible for the emergence of interference effects." Clearly, the emphasis is here on the employment of probability *amplitudes*. See also Moreira and Wichert [13].

The joint distribution and marginalization formula now works with the quantum-like approach via the use of probability amplitudes.

The paper provides for a useful comparison between classical and quantum-like Bayesian networks and the authors report that (p. 23) "The quantum-like model achieved a mean error of 17.86 percent compared with the 13.78 percent mean error obtained in a classical setting. The general results show that the average error over the 19 random variables was, in scenarios where nothing is observed, for the quantum-like Bayesian network 5.90 percent compared to a 22.85 percent error in the classical network."

7.5 Quantum Counting and Open Systems in Social Science

7.5.1 Quantum Counting

Fabio Bagarello in [14] pioneered a new approach based on the so-called number operator to population methods. The number operator, widely used in quantum theory, can also be used to model discrete populations in social science and ecological contexts. In work by Robinson et al. [15] we argued that because natural numbers (as numbers used in counting) can represent the spectrum of the eigenvalues of an operator, one obtains an operator-valued calculus which can be used in other avenues than physics. The basics in this approach start with the equation:

$$\hat{\Gamma}|\gamma\rangle = \gamma|\gamma\rangle, \tag{7.5}$$

where $\hat{\Gamma}$ is an operator, operating on a state, $|\gamma\rangle$ that has an eigenvalue γ, giving some information about the state. The key issue we remarked in the paper was that a system which allows a mathematical representation is all that is needed. Hence, the broad scope of applicability to areas well outside of physics. In comparison with population predator–prey models like the well-known Lotka–Volterra model, this approach shows that the population interactions are now modelled via Hamiltonian operators.

7.5.2 Open System

The so-called open systems have found applications in political science (see Khrennikova and Haven, [16]) and we can also think of such systems in finance. In political science, the open-system approach lends itself well to formalizing the bath of information (see Khrennikova et al., [17]). Typically in open systems we shall find recourse to the so-called GKSL-equation which can be seen as a Markovian quantum master equation. The GKSL-equation can transform pure states to mixed states.

When we introduce the idea of a 'mental bath' we think of a financial decision-making environment where we focus on information adaptive behaviour. One may claim that financial agents centre their decision-making behaviour around information sourced from other agents and the environment. In this research vein Fabio Bagarello et al. in [18] applied the formalism of quantum field theory. It is essentially the information space which is mathematically expressed by quantum fields.

As we remarked, in the paper by Fabio Bagarello et al., [18] , such adaptive systems can be modelled via open systems, where essentially the state of an agent is extracted from the general state of a system which is formed by the agent and his/her environment. As we mention in that work, the dynamics are reduced to the state space of this agent. When the environment is represented as a quantum field, we essentially follow the work which was laid out by Bagarello in [19] .

In a nutshell, a number operator for a market participant or player is defined and it will be the dynamics to that operator which is the focus of this discussion. The dynamics will be dependent on the specification of a Hamiltonian – but for an open system! The importance of the open system is that agents really explicitly interact with their environment. The full Hamiltonian is quite involved and consists of three parts:

$$H = H_0 + H_I + H_{int}, \tag{7.6}$$

where:

1. H_0 is a *free* Hamiltonian,
2. H_I describes the interaction between the players and their neural systems, and
3. H_{int} describes two different interactions between two players.

We note that

$$H_{int} = \mu_{ex}\left(b_1^\dagger b_2 + b_2^\dagger b_1\right) + \mu_{coop}\left(b_1^\dagger b_2^\dagger + b_2 b_1\right), \tag{7.7}$$

where $\hat{n}_j = b_j^\dagger b_j$ is the number operator of the jth player. The operators b_j also have a simple matrix form. All of μ_{ex} and μ_{coop} are real quantities. Modulating those two parameters leads to a framework where players can cooperate or not. The quantum field approach does lead to interesting results even within an extremely complex framework which here is steered by a very complex Hamiltonian H. The approach followed here tries to simplify a very difficult problem – namely an open system which is highly different from the isolated system in quantum mechanics as portrayed by the use of the Schrödinger PDE.

7.6 Non-Hermitian Hamiltonians in Finance

In Belal Baaquie's work, the main driver will be quantum mathematics, without a necessary reference to physics. Belal Baaquie in [20] (p. 1666) did mention in earlier work that "the interpretation of the symbols of quantum mechanics in such disciplines have no fixed prescription"

Within the setting of option pricing, Baaquie's approach revolves around the idea that the option pricing function is an observable wave function. The Black–Scholes Hamiltonian is non-Hermitian and it is of the form in [20] (p. 1668):

$$H_{BS} = \left[-\frac{\sigma^2}{2} + \left(\frac{\sigma^2}{2} - r \right) + r \right] \exp(x) = 0, \tag{7.8}$$

where σ is stock price volatility; r is the risk-free rate (a rate of interest which is observable by all market participants); and x is the growth rate of stock S. The reference to using quantum mathematics and not involving the physics of quantum mechanics relates narrowly to the fact that H_{BS} is *not Hermitian* and, as we said before, this non-Hermiticity occurs because of the nonarbitrage constraint[4].

The Schrödinger PDE expressing the time-dependent evolution of the wave function requires a Hamiltonian which is Hermitian. For open systems, we can steer away from using this PDE since it is valid for only isolated systems. As we have remarked in other work (see Haven et al. in [21]) the non-Hermiticity leads to very difficult problems such as (i) energies are expressed with complex numbers and (ii) conservation of probability is violated.

However, as is known to some extent, non-Hermitian Hamiltonians can be considered to deliver real eigenvalues, but only in special cases. We remarked in Haven et al., [21] that real physical phenomena with non-Hermitian Hamiltonians can be shown to exist with the following simple example (see Bender and Boettcher, [22], and also Bender, [23],

$$H = p^2 + x^2(ix)^{\epsilon}, \tag{7.9}$$

where $\epsilon > 0$ is a real number. This Hamiltonian has real (positive) energy levels (when ϵ is positive) and probability is conserved.

7.7 Quo Vadis: Expanding Mathematical Modelling?

A question of possible interest may be as follows: given the quantum potential contains the amplitude function of the polar form of the wave function, could we pose the 'relevant' query, what 'reality' does the wave function refer to? After all, we try to identify with the quantum potential that it captures a particular angle

[4] It is notoriously difficult to come up with a unique option price if arbitrage is allowed.

of public information. In Ringbauer et al., [24], the question is asked whether the wavefunction corresponds to the underlying reality, *or,* does the wave function only represent partial knowledge? This is clearly a difficult question to answer.

It should not come as a surprise that a parameter-rich formalism which aids in the modelling of 'information' in economics/finance is a very welcome addition to modelling in those fields. In effect, as one could claim, information is a major player in driving prices. As we saw, for higher moment probability density functions, the quantum potential may well be a useful formal device to refer to when considering information from those moments. In that respect, the example of the classical potential we considered does not really answer this need for a formalism to model information.

We could distinguish at least three types of models which allow the modelling of information:

- A first model: which is a 'No memory' model with asymmetry between past and future time,
- A second model: which is a 'No memory' model with symmetry between past and future time, and
- A third model: which is a memory model.

We commonly use in finance a version of Brownian motion which allows for the lognormal density of price returns. Such Brownian motion has so-called white noise that is, it does not allow for memory between price occurrences. Brownian motions with a so-called 'Hurst exponent' allow for the introduction of memory. This stochastic differential equation was popular at some point in finance (see Rodgers [25]). 'Physicsworld' in 2011 reported (see Commissariat, [26]) that the usual assumption that kicks are random when a particle descends in, say, a water vessel, may not always be true. Nonrandomness occurs when the densities of the particle and fluid are similar. This gives rise to the idea of so-called 'coloured noise'.

7.7.1 A First Model: 'No Memory' with Asymmetry between Past and Future Time. A Fisher-Type Information Model?

A first model, which is certainly extraneous to the economic literature, is the model proposed in this section. The model here does show a possible connection with the Hawkins and Frieden approach we discussed earlier. Edward Nelson, [27], who did important work in the stochastic approach to quantum mechanics, assumes a Brownian motion:

$$dx(t) = b_{+}(x,t)dt + \sigma dW(t), \tag{7.10}$$

where $dW(t)$ is a Wiener process; σ is the diffusion coefficient; and $b_+(x,t)$ is the drift function. Since $E(dW(t)) = 0$, one can define

$$D_+x(t) = \lim_{\epsilon\to 0} E\left[\frac{x_{n+1} - x_n}{\epsilon}\right] = b_+(x,t). \tag{7.11}$$

Similarly, he assumes a Brownian motion of the following type:

$$dx(t) = b_-(x,t)dt + \sigma dW(t) \tag{7.12}$$

and one can then define

$$D_-x(t) = \lim_{\epsilon\to 0} E\left[\frac{x_n - x_{n-1}}{\epsilon}\right] = b_-(x,t). \tag{7.13}$$

Two velocity fields are then defined:

- mean velocity:

$$v(x,t) = \frac{1}{2}(b_+(x,t) + b_-(x,t)), \text{ and} \tag{7.14}$$

- osmotic velocity:

$$u(x,t) = \frac{1}{2}(b_+(x,t) - b_-(x,t)). \tag{7.15}$$

We know that in Newtonian mechanics $b_+(x,t) = b_-(x,t)$, and hence $u(x,t) = 0$. We have asymmetry between past and future time, when $u(x,t) \neq 0$. Edward Nelson has already, in some sense, made allusion to the existence of Fisher information (or Fisher-type information). He defines

$$u(x,t) = \nabla \ln p(x,t), \tag{7.16}$$

where $p(x,t)$ is the probability density function. In that sense, can we say that $u(x,t)$ is narrowly related to Fisher information:

$$I \equiv \left\langle \left[\frac{\partial \ln p}{\partial x}\right]^2 \right\rangle. \tag{7.17}$$

If, for instance,

$$p(x,t) = \exp(2R(x,t)), \tag{7.18}$$

then

$$u(x,t) = \sigma^2 \nabla R(x,t) \tag{7.19}$$

and the quantum potential (omitting mass and Planck constant) is

$$\frac{1}{R}\nabla^2 R. \tag{7.20}$$

What is important to note here is that without this asymmetry between past and future, we cannot make this link to Fisher information. The model is also memory-free, since white noise is at the source of randomness.

7.7.2 A Second Model: 'No Memory' with Symmetry between Past and Future Time

We briefly mentioned before the approach by Detemple and Rindisbacher, [28], relative to the modelling of private information. Their notion of 'Private Information Price of Risk (PIPR)' is defined: as representing the changed price of risk when private information becomes available (p. 190). They define an enlarged filtration to accommodate this private information. The information gain (public information with the addition of private information) can be measured with an appropriate entropy measure. The full model occurs within a memoryless (white noise) environment and there is no asymmetry between past and future time.

The increased filtration, that is, the additional information does NOT affect the nonzero quadratic variation. This is a large difference relative to the first model, since there, when accounting for additional information, zero quadratic variation obtains.

7.7.3 A Third Model: A 'Memory' Model

This model is close to the Bohmian mechanics–hydrodynamics approach we discussed earlier when we laid out some of the basic building blocks from the paper by John Bush, [29], (and see also Eddi, [30]; Fort and Eddi, [31]; Couder and Fort, [32]; and also Hardesty, [33]) but there is now the additional dimension of memory. The model looks at a so-called 'walking droplet'. There seems to be an explicit 'context'[5] here – and it is definable as follows.

- Preparation is necessary **before** time $t = 0$: the droplet bounces on a surface which needs to vibrate in a certain way (otherwise, droplets will coalesce with the liquid they drop on. The droplet is made out of the same liquid).
- Preparation affects the trajectory on $t = 0$ and following: the damping time of the wave formed by the impact of the droplet on the vibrating surface is dependent on the vibration frequency distance from a certain threshold.

We can ask ourselves here the question whether we have 'context conditioning' versus 'event conditioning' (based on Bayes's formula)? Prima facie, we can think of the varying vibration frequency of the bath over time as equivalent to saying that the global picture is not inconsistent with the constituent pictures and in that

[5] We need to be careful with the use of the word 'context' here, as it also has a very precise meaning in quantum mechanics.

sense, maybe we can claim that context in the 'quantum' sense may not exist here?[6] In this macroscopic model, the so-called 'walker' exhibits wave–particle duality. In this macroscopic version of the double-slit experiment, the walker droplet passes through one slit or the other. The guiding waves pass through both slits. Hence, the walker droplet 'feels' the second slit by virtue of the pilot wave (see Bush, [29]). One can thus say that in this macroscopic fluid bath, the motion of a droplet is driven by an interaction with a superposition of waves emitted by the points the droplet has visited previously. It is in that sense that we can say there is memory. It has been remarked by Fort et al., [31], that the path memory we mentioned may be akin to some sort of temporal nonlocality.

7.7.4 Some Social Science Examples Which May Employ Elements of Model 3

We provide here some heuristic examples on how the macroscopic fluid model could provide for potentially useful parameters in a social science setting.

- **Example 1:**

The time duration of the information impact (damping time, τ) is dependent on the context (i.e. preparation: the fluid frequency in the bath is close to or far from the so-called 'Faraday threshold'): *by analogy – into social science*: why do IPOs[7] and their ensuing price level (height) have such differing impact before or after an election?

- **Example 2:**

Past bounces of the droplet affect the current trajectory: memory is important *by analogy – into social science:* in mathematical finance this is resisted: past information is not informative of the future.

- **Example 3:**

The gradient of the height, ∇h, determines the surface slope (and will determine the way the droplet bounces) *by analogy – into social science*: in so-called technical analysis, a detailed analysis of past price increments can be used to chart future price behaviour. This idea goes counter to the academic finance stance, which says that no prediction is possible via the concept of martingale. In summary, the parameter richness of model 3 can be seen as promising and we should ask ourselves what the hard data-driven analogies could be on parameters like damping time (τ), height ($h(\tau)$), and ∇h.

[6] The reader is referred to some deep ideas Samson Abramsky is bringing on the meaning of context. Please see for instance [34].

[7] IPO: Initial Public Offering. Shares of a company which go public are offered for sale to institutional investors and other types of investors.

7.8 Quo Vadis: Limitations of Mathematical Modelling?

In recent work, Arkady Plotnitsky [35] probed the limitations of quantum-like models. We hope this short discussion can be seen as an interesting way to close this chapter of the book. Although there is disagreement in the current literature on the applications of quantum formalisms to areas outside of quantum physics as to whether the 'quantum-like' approach is the unique stance in that research, we shall limit this short expansion of ideas here to quantum-like approaches only.

Arkady Plotnitsky makes the point that the effectiveness of mathematics in the hard sciences is very probably at much higher levels than it is in the social sciences. We, of course, can come up with some immediate arguments, such as the complex (and difficult-to-define) framework we work in the social sciences as opposed to the hard sciences. Related to this though is an argument that Plotnitsky makes which shows the parallels between the process of generating quantum physics (in physics) and the process which can explain the creation (and use) of quantum probability in (social science–based) decision-making. Plotnitsky, [35], (p. 2) says: "While physically described by classical physics, the experimental data obtained in quantum experiments cannot be predicted by classical physics. This incapacity led to the rise of quantum theory, from M. Planck's discovery of the black body radiation law in 1900 on, eventually leading to QM and QFT. Beginning with A. Tversky and D. Kahneman's pioneering work in the 1970s to the 1980s (e.g. Tversky and Kahneman [36], [37]), it has been the presence of probabilistic data akin to those of quantum physics that suggested using quantum-like models (e.g. Haven and Khrennikov, [38], Busemeyer and Bruza, [39])." One could make the claim that this direction of thought may well be akin to what Arkady Plotnitsky mentions when the mathematical edifice of quantum mechanics was set up, [35], p. 4: "The mathematical features of the formalism were inferred (surmised and sometimes guessed), by Heisenberg from the physical features of quantum phenomena." And furthermore, he writes (p. 4) "The probability amplitude is $\langle\lambda_i|\psi\rangle$ (λ_i is an eigen-value and ψ is the wave function), and the corresponding probability is $p_i = \langle\lambda_i|\psi\rangle^2$. In this view, quantum amplitudes and linear superpositions do not represent anything physical, unlike in classical physics where both represent physical processes." This is an interesting construction which we think in some sense does follow the way we use the quantum formalism in social science. From the outset in most of what is being done in formal setups in social science, the 'social science' process representation of formal mathematical concepts is virtually absent. As an example, we cannot think of an oscilloscopic interpretation of a time-dependent price dynamic. As another example, Brownian motion whilst very well argued for in physics, is essentially of metaphorical use in finance.

The next question Arkady Plotnitsky then asks is how the use of those quantum-like models may then reveal any assumptions as to whether one thinks our brain may be quantum mechanical in nature. Plotnitsky, [35], p. 3, says, "How the physics of the brain makes thinking or consciousness, as we experience it, remains an unanswered question, sometimes referred to as 'the hard problem of consciousness' (Chalmers, [40])." Somewhat related to this point is also the distance we may want to take from the idea that what we observe may in the end influence the phenomenon. Surely the brain – if quantum mechanical – may enter into some permanent loop of the observe–influence issue. As is well known, the observe–influence problem is a nonissue in classical physics, but it is a big challenge in quantum mechanics. Many social sciences processes may not escape the observe–influence problem. As we have documented in the book, decision-making problems are squarely within this problem setting. Question order reversals and the impact such reversals have on answers to test subjects is an example. However, other areas of social science may well not be subject of this observe–influence problem.

Here, however, we may wonder: if indeed there is some parallel semblance (in relation to the birth of quantum physics) of theoretical construction toward quantum-like, there are still very large differences. For instance, Plotnitsky makes the very interesting remark that, [35], p. 7, "In the case of human experience, there are no entities analogous to quantum objects, such as electrons or photons, whose physical identity could be maintained in quantum experiments. The observed states of quantum objects are different, too. However, while mental representations can have parts shared between different subjects, they are ultimately different for each subject. In quantum physics, a measured physical state of, say, an electron is always that of an electron."

Finally, as we have been hinting in this chapter, the existence of open systems in attempting to use quantum-like real-world needs to be kept in mind. This is very consequential in that such modelling entails the use of approaches like GSKL, thus, away from the usual Schrödinger setting.

References

[1] Baaquie, B. E. (2007). *Quantum Finance: Path Integrals and Hamiltonians for Options and Interest Rates*. Cambridge, UK: Cambridge University Press.
[2] Orrell, D. (2021). The color of money: threshold effects in quantum economics. *Quantum Reports*, 3, 325–332.
[3] Sarkissian, J. (2020). Quantum coupled-wave theory of price formation in financial markets: Price measurement, dynamics and ergodicity. *Physica A*, 554, 124300.
[4] Orrell, D. (2020). A quantum model of supply and demand. *Physica A*, 539, 122928.
[5] Focardi, S., Fabozzi, F. Mazza D. (2020). Quantum option pricing and quantum finance. *Journal of Derivatives*, 28 (1), 79–98.
[6] Segal, W., Segal, I. (1998). The Black–Scholes pricing formula in the quantum context. *Proceedings of the National Academy of Sciences*, 95 (7), 4072–4075.
[7] Accardi, L., Boukas, A. (2007). The quantum Black–Scholes equation. *Global Journal of Pure and Applied Mathematics*, 2 (2), 155–170.

[8] Haven, E., Khrennikova, P. (2018). A quantum-probabilistic paradigm: non-consequential reasoning and state dependence in investment choice. *Journal of Mathematical Economics*, 78, 186–197.
[9] Tversky, A., Shafir, E. (1992). The disjunction effect in choice under uncertainty. *Psychological Science*, 3 (5), 305–309.
[10] Croson, R. (1999). The disjunction effect and reason-based choice. *Organizational Behavior and Human Decision Processes*, 80 (2), 118–133.
[11] Moreira, C., Haven, E., Sozzo, S., Wichert, A. (2018). Process mining with real world financial loan applications: improving inference on incomplete event logs. *PLOS One*, 13 (12), e0207806.
[12] van der Aalst, W., Weijters, T. and Maruster, L. (2004). Workflow mining: discovering process models from event logs. *IEEE Transactions on Knowledge and Data Engineering*, 16, 1128–1142. https://doi.org/10.1109/ TKDE.2004.47
[13] Moreira, C., and Wichert, A. (2016). Quantum-like Bayesian networks for modeling decision making. *Frontiers in Psychology*, 7 (11). https://doi.org/10.3389/fpsyg.2016.00011.
[14] Bagarello, F. (2013). *Quantum Dynamics for Classical Systems*. New York, NY: John Wiley & Sons.
[15] Robinson, T., Haven, E. and Fry, A. (2017). Quantum counting: Operator methods for discrete population dynamics with applications to cell division. *Progress in Biophysics and Molecular Biology*, 130, 106–119.
[16] Khrennikova, P. and Haven, E., (2015). Instability of political preferences and the role of mass media: A dynamical representation in a quantum framework. *Philosophical Transactions of the Royal Society A*. 374, 20150106.
[17] Khrennikova, P., Haven, E. and Khrennikov, A. (2014). An application of the theory of open quantum systems to model the dynamics of party governance in the US political system. *International Journal of Theoretical Physics*, 53 (4), 1346–1360.
[18] Bagarello F., Haven E. and Khrennikov A. (2017). A model of adaptive decision-making from representation of information environment by quantum fields. *Philosophical Transactions of the Royal Society A*, 375, 20170162.
[19] Bagarello, F. (2012). *Quantum Dynamics for Classical Systems: with Applications of the Number Operator*. New York, NY: John Wiley & Sons.
[20] Baaquie, B. (2013). Financial modelling and quantum mathematics. *Computers and Mathematics with Applications*, 65, 1665–1673.
[21] Haven, E., Khrennikov, A. and Robinson, T. (2017). *Quantum Methods in Social Science: a First Course*. Singapore: World Scientific Publishers.
[22] Bender, C. M., Boettcher, S. (1998). Real spectra in non-Hermitian Hamiltonians having PT symmetry. *Physical Review Letters*, 80, 5243.
[23] Bender, C. M. (2016). PT symmetry in quantum physics: From a mathematical curiosity to optical experiments. *Europhysics News*, 47, N 2, 18–20.
[24] Ringbauer et al. (2015). Measurements on the reality of the wavefunction. *Nature*, 11, 249–254 https://doi.org/10.1038/NPHYS3233.
[25] Rodgers, L. C. G. (1997). Arbitrage with fractional Brownian motion. *Mathematical Finance*, 7 (1), 95–105.
[26] Commissariat, T. (2011). New twist on Brownian motion seen for the first time. *PhysicsWorld*. Philadelphia, PA: IoP Publishing.
[27] Nelson, E. (1966). Derivation of the Schrödinger equation from Newtonian mechanics. *Physical Review*, 150, 1079–1085.

[28] Detemple, J., Rindisbacher, M. (2015). The private information of risk. Pages 190–213 in *The Palgrave Handbook of Post Crisis Financial Modelling* (Eds. Haven, E. et al.). London: Palgrave-MacMillan Publishers.
[29] Bush, J. W. M. (2015). Pilot wave hydrodynamics. *Annual Review of Fluid Mechanics*, 47, 269–292.
[30] Eddi, A. et al. (2011). Information stored in Faraday waves: the origin of a path memory. *Journal of Fluid Mechanics*, 674, 433–463.
[31] Fort, E., Eddi, A. et al. (2010). Path-memory induced quantization of classical orbits. *Proceedings of the National Academy of Sciences of the USA*, 107 (41), 17515–17520.
[32] Couder, Y., Fort, E. *Discussion on macroscopic pilot wave system.* see: http://math.mit.edu/~bush/?p=2984.
[33] Hardesty, L (2014). *New math and quantum mechanics: fluid mechanics suggests alternatives to quantum orthodoxy.* M.I.T. News: see https://news.mit.edu/2014/fluid-systems-quantum-mechanics-0912.
[34] Abramsky, S., R. S. Barbosa, K. Kishida, R. Lal and S. Mansfield. *Contextuality, cohomology and paradox.* https://logic.berkeley.edu.
[35] Plotnitsky, A. (2022). Most tantumising state of affairs: mathematical and non-mathematical in quantum-like understanding of thinking. *Frontiers in Psychology*, 13: 934776.
[36] Tversky, A., Kahneman, D. (1973). Availability: a heuristic for judging frequency and probability. *Cognitive Psychology*, 5, 207–237. 90033 doi: https://doi.org/10.1016/0010-0285(73)90033-9.
[37] Tversky, A., Kahneman, D. (1983). Extensional versus intuitive reasoning: the conjunction fallacy in probability judgment. *Psychological Review*, 90, 293–315. doi: 10.1037/0033-295X.90.4.293.
[38] Haven, E., Khrennikov, A. (2013). *Quantum Social Science.* Cambridge: Cambridge University Press. doi: 10.1017/CBO9781139003261.
[39] Busemeyer, J., Bruza, P. D. (2014). *Quantum Models of Cognition and Decision.* Cambridge: Cambridge University Press.
[40] Chalmers, D. (2010). *The Character of Consciousness.* Oxford: Oxford University Press. doi: 10.1093/acprof:oso/9780195311105.001.0001.

Part III

Quantum Physics-like Modelling

8

Motivations and Foundations

The quantum-like research effort has mostly concentrated on using the formalism of quantum mechanics to applications outside of quantum physics, specifically in decision-making. Thus, the emphasis has been, in most of the successful applications, on generalizing probability and then coming to strong results like that with the QQ equality (and its generalization with quantum instruments) as discussed in Chapter 5. The movement toward quantum physics–like models is a subtle one – as the reader will experience from reading this chapter. Not only is it subtle but it is also very new within the historical timeline of the movement which applies quantum physics to areas outside of its remit. Besides the important notion (to which we have already alluded in Chapter 2) of an ensemble of identical systems, in this chapter we shall also be concerned about the quantum physics–like interpretation of the wave function in the social scientific context. This is a difficult task, but if properly tackled, it could lead to moving the full quantum-like movement even more forward.

In Part II we took a review of several important contributions exemplifying the application of quantum theory in the social sciences. These contributions provide models – of human behaviour and the resulting (decision-making / economics-finance) activities over a broad variety of contexts – developed by applying the mathematical formalism of quantum theory. Such developments can be regarded as the outcomes of the use of the language of *quantum mathematics* in the social sciences [1]. Such a language originates from the mathematical foundations of quantum theory, which include topics such as operator algebras, linear vector spaces and the theory of probability.

With hindsight, it can be remarked that the corresponding research efforts seem to be based on the tacit hypothesis that the mathematical language used for describing, comprehending, and analyzing the quantum physical phenomena can work analogously in the social scientific domain as well. We prefer to call the resulting quantum-like models the quantum *math*–like models, in the sense that they provide quantum math–analogous comprehension and analyses of the social-economics-finance contexts.

However, given several other important aspects of quantum physics, we need not restrict ourselves only to the development of the quantum math–like models of the human activities in the economics/finance contexts. On the contrary, encouraged by the success of the developments discussed in Part II, we proceed to explore how such a research campaign could be evolved further, by raising the following questions: how could quantum *physics*–analogous – as opposed to quantum *math*–analogous – models be constructed for economics and finance? How would such models differ from the quantum math–like ones? Would these models offer an advantage over the quantum math–like models? The purpose of the present chapter and the remaining chapters of this book is to provide answers to these questions. In the following we briefly explain the rationale behind our exploration in the present part.

Let us begin by recalling that quantum theory is, primarily, a theory of physics. The quantities involved in any theory of physics always have well-defined, definite physical meanings, which dictate the overall interpretation and meaning of the theory. Somehow, the problem of interpretation is much more challenging in the case of quantum theory than that in the case of the previous physical theories. Indeed, physicists have come up with a variety of altogether different interpretations of quantum theory (and especially, of the wave function) without much disagreeing with each other on its *mathematical* foundations. Moreover, these interpretations have given rise to several different (and debatable) perspectives on the fundamental nature of physical reality (see Subsection 1.1.4, and also Subsection 8.1.3), which in turn, have led to seemingly never-ending dialogues at the interface of the epistemology and the ontology of microscopic physics.

For the construction of quantum physics–analogous models what we require is, thus, a 'quantum physical foundations focussed' approach. As explained in Sections 8.1 and 8.2, such an approach will be seen as more exhaustive than that employed in the construction of the quantum math–analogous models. One of the central features of such an approach is its insistence on developing the economics/finance analogues of the various aspects of the physical foundations of quantum theory in addition to using the mathematical language of quantum theory. This way, the approach will lead to the construction of new quantum models that will be able to provide physics-analogous interpretations of the happenings of the economics/finance world (Chapters 9 and 10). It could thus help expand the variety of quantum-like models for economics and finance.

One may argue that the *physical interpretation* of any theory of physics is communicated by its *mathematical language* itself and hence, there is no need to worry about the former separately if the latter has already been used. However, we beg to differ and argue that the mathematical expressions from physics themselves cannot imply or indicate or give rise to any physical interpretations of the quantities involved in those expressions. Our argument can be seen to share its spirit with what Feynman says (see [2] ch. 2, sec. 2.1), for we also think that a pure mathematical understanding of any aspect of a theory of physics may not guarantee an understanding of the *physics* of that aspect[1].

Having said this, we do not aim to evaluate the already developed quantum-like models from the point of view of quantum foundations and interpretations. Rather, as remarked earlier, we prefer to explore the construction of new models by paying serious attention to the physical foundations of quantum theory. Our exploration begins with a brief overview of the physical foundations of quantum theory (Section 8.1), which will turn out to be helpful in deciding the criteria for a quantum economics/finance model to qualify as a quantum physics–analogous model and therefore, will lead to the general expectations from such models (Section 8.2).

As we will see, the new approach will require framing the ideas, concepts and notions – to be used for expressing the happenings of the economics/finance world – having precise analogies with those expressing the happenings of the physical world. In other words, our approach will enforce a systematic development of 'analogies' between quantum physics and the economic sciences by viewing it as an essential task. For this we follow some proposals from the philosophy of analogical reasoning.

The philosophy of analogical reasoning and analogical arguments is a well-established area of academic research belonging to the parent field of philosophy of science. And, as a prerequisite for the preceding task, a certain level of comprehension of this area will be required. In Section 8.3 we give an overview of the same. There, among other things, we discuss briefly four different theories that provide guidelines for developing analogies between two domains and also, some specific criteria for evaluating analogical arguments. These will be Aristotle's theory (as perhaps the most ancient theory of analogy); Hesse's theory (emphasizing the 'material' aspect of analogy); Gentner's structure-mapping theory (emphasizing the importance of structural isomorphism and the so-called 'systematicity'); and Bartha's articulation model. The philosophical insights offered by these theories lead us to a certain strategy for constructing quantum physics–analogous models (Section 8.4). We implement this strategy in the next two chapters (Chapters 9 and 10) to model a specific finance context.

1 We come back to this point in Section 8.4.

8.1 On the 'Physical' Foundations of Quantum Theory

What are the *physical*, as opposed to the *mathematical* foundations of any theory from physics? For instance, what could be considered as the physical foundations of the theory of classical electrodynamics? The physical foundations of classical electrodynamics lie in the notions of charge, current, electric–magnetic fields and so on. And, these foundations become relevant when it comes to an understanding of the processes, described in terms of these notions, in an un-mathematical, qualitative manner.

To see how this works, we refer the reader to [2], where, on many occasions, Feynman's treatment of the subject of electromagnetism helps develop a physical, intuitive comprehension of the processes involving charges, currents and electric–magnetic fields.[2]

Physical foundations of a theory become apparent also in the experimental observations that lead to the formulation of the theory. In the case of electromagnetism, for example, the experiments performed by Oersted and Faraday did allow an intuitive grasp of the connection between the electric and the magnetic fields, without necessitating the prior knowledge of the 'formal' mathematical language of electromagnetism, which is provided by vector calculus.

Of course, for any theory of physics, its physical foundations go hand in hand with the mathematical language used in the theory. This seems particularly true for the fundamental theories of nature developed after the scientific revolution. Such theories demonstrate an inseparable association of the mathematical preciseness with the natural philosophy[3]. In fact, such an inseparability makes it difficult to delineate the physical foundations of any theory of physics. However, this does not mean that it is impossible to appreciate their distinct and vital role in any theory of physics, and, for the purpose mentioned earlier, we are required to portray them in the clearest possible manner.

Here, we would like to refer again to [2, ch. 2, sec. 2.1] where Feynman considers 'physical understanding' as un-mathematical, imprecise and inexact. Also, he asserts that, in spite of having such (pessimistic-sounding!) features, a physical understanding is absolutely necessary for a physicist. We think that somehow Feynman has already hinted at a certain distinction between the *physical*, as opposed to the *mathematical* understanding of a theory of physics. And, we hope that, in a

[2] Feynman's intuitive view was not only seen in the manner he taught physics, but also in the way he pursued research in theoretical physics. Julien Schwinger, who shared the 1965 Nobel Prize in Physics with Feynman and Tomonaga, regards Feynman as the *outstanding intuitionist*, see [3]. It is interesting to note that, 'intuitionism' asserts that mathematical knowledge essentially involves intuition and mental construction. This way, it tends to reject the notion of independent mathematical objects. For details, see, for example, [4].

[3] Recall the discussions from Section 1.1, where we viewed the scientific revolution as a process by which the ancient 'natural philosophy' version of physics evolved into its modernized 'natural science' version.

similar manner, it is possible to appreciate the distinction between the physical and the mathematical foundations of a theory from physics, such as quantum theory.

Starting from John von Neumann's classic text [5] with a mathematically rigorous treatment of quantum mechanics, today we have many excellent and detailed accounts devoted mainly to the *mathematical* foundations of quantum theory, see, for example, [6]. Besides these, there are equally excellent accounts of the quantum foundations (e.g. [7]), which predominantly provide physics-focussed discussions, for example, of the issues such as the measurement problem or the problem of classical-to-quantum transition or the problem of wave function interpretation. Yet, it is not a common practice to attach the adjective 'physical' to the foundations such accounts discuss!

Similar to the inherent impreciseness – as remarked by Feynman – in a pure physical understanding of a given situation, we think that there is an unavoidable vagueness when it comes to the matter of focussing on and spelling out the pure physical foundations of a theory. Still, by following Feynman's remark, we may assert that a certain clarity of the meaning of physical foundations of quantum theory will nevertheless be absolutely necessary, if the aim is to construct quantum *physics–analogous* models for economics and finance.

But, what could be regarded as the physical foundations of quantum theory? With reference to the preceding discussion, the reader might have already anticipated that the physical foundations would refer mainly to the nonmathematical aspects of quantum theory. This is in the sense that such foundations would neither directly pertain to nor essentially involve the details of the (generally accepted) quantum mathematical language. In other words, the aspects of physical foundations could be described without much getting into the intricacies of the 'pure mathematical' structure of quantum theory; and, by the same token, the mere knowledge of the 'pure mathematical' details would not help a comprehension of these foundations. In the following, an attempt is made to describe such foundations in a particular manner.

The physical foundations of quantum theory can be regarded as having theoretical, experimental and philosophical aspects. Somehow, all these aspects can be seen to lie at the classical–quantum interface. For instance, the quantum physical foundations could be considered as having to do with the following:

- the overlaps between the dynamical formulations of the classical and the quantum theories, which can be seen in the similarities between the treatments, usage, and interpretations of the kinematic and dynamic physical quantities that are common to both the formulations;
- the similarity between the nature of the observable outcomes in the spectroscopic and interferometric experiments – whether the experiment belongs to the classical domain or the quantum domain; and

- the room the quantum theory allows for a variety of interpretations of the wave function and the nature of *physical* reality that it implies, which seems to be much more intricate than what classical theory implies.

In the following subsections we elaborate on each of these aspects one by one. As we will see in the next section (8.2), these aspects lead us to a clear idea of what can be expected from the 'quantum physical foundations focussed' approach and hence, from the resulting quantum physics–analogous models for economics and finance.

8.1.1 Classical–Quantum Overlaps

We already saw in Section 2.4 that there exists a subtle connection between the classical Hamilton–Jacobi equation and the quantum Schrödinger equation. Indeed, being one of the stages of the developments in the many-centuries-old discipline of physics, the seeds and the roots of the formulation of quantum theory lie in the preceding (i.e. classical) physics paradigms [8].

The classical physics paradigm represents a set of theories that includes, for example, the Newtonian mechanics, the Lagrangian–Hamiltonian–Jacobian analytical mechanics, statistical mechanics and thermodynamics, optics and electrodynamics, classical field theories and relativity. The quantum paradigm involves a set of theories such as quantum mechanics, quantum statistical mechanics, quantum optics, which, in general, are the extensions of their respective classical counterparts applicable in the 'micro' domain. Therefore, it is not surprising that the formulation of quantum theory involves several ideas and notions that have their origins in the formulation of classical theories.

In the following we give a (far-from-complete) list of assorted items. Each one of these items indicates, in some way or another, an overlap between the classical and quantum theories:

- To explain the experimentally observed blackbody spectrum, Max Planck modified the 'classical' Rayleigh–Jeans theory. He thus resolved the issue of ultraviolet catastrophe, that is, the classical prediction that the energy emitted by an ideal black body at thermal equilibrium would tend to infinity in the short-wavelength limit. In the high-wavelength limit, there is an agreement between the predictions of the Planck and the Rayleigh–Jeans theories, whereas their disagreement is much more pronounced for shorter and shorter wavelengths.
- In his celebrated 1900 derivation, Planck introduced the universal constant, the *quantum of action*, 'h' and obtained the correct expression for the intensity (and hence, the energy) distribution function of cavity electromagnetic radiation, at a given temperature, across a range of wavelengths. The interpretation of Planck's constant as the quantum of action points to a well-founded *classical* notion of action.

- In 1905 Albert Einstein provided an explanation of the photoelectric effect by making use of Planck's 'quantum' hypothesis. For this, he gave a simple relation, namely, $K_{max} = h\nu - W$, where ν is the frequency of the incident electromagnetic radiation, K_{max} is the maximum kinetic energy with which the electrons can leave their atomic binding and $W = h\nu_0$ (the work function) is the minimum energy required to remove an electron from the surface of the material, with ν_0 as the threshold frequency – that is, the minimum frequency required to ensure adequate incident electromagnetic energy in the form of individual photons. This relation leads to a *quantum* theoretical explanation of the observed properties of the photoelectric effect. Yet it involves notions such as kinetic energy and radiation frequency whose meanings are originally founded in the *classical* physics.
- The Bohr atom model – which proposed a Solar system–like structure of hydrogen atom – represents a combination of *classical* notions (electron–proton electrostatic interaction, angular momentum) and *quantum* ideas (e.g. discrete spectrum of the allowed energies for the orbiting electron, quantization of angular momentum).
- In the modern formulation of quantum theory[4], dynamical variables, such as momentum, kinetic and potential energies, angular momentum and so on are viewed as operations on the system states in quantum mechanics. However, the variables themselves have their conceptual and interpretational origins in the theory of classical mechanics[5].
- The ideas of the inherent properties of physical systems, such as mass and charge, having origins in classical physics, continue to play their roles in quantum physical contexts, with the same original physical meanings. In other words, whether appearing in the expression of a classical Hamiltonian function or in the expression of a quantum Hamiltonian operator, the physical interpretations of mass and charge are the same.
- The idea of an ensemble, playing an important role in quantum physics, has its origins founded in the formulation of classical statistical mechanics.

We invite the reader to add more items in the preceding list of classical-quantum overlaps!

Thus there are many ideas and notions that are used in the modern formulation of quantum theory, but their physical interpretations and meanings are well-founded originally in the classical physics.

[4] Here, by 'modern formulation' we mean the formulation that began with the de Broglie hypothesis.
[5] The quantum idea of intrinsic 'spin' of particles is, of course, the notable exception.

8.1.2 From 'Intensity' to 'Probability'

Just like any other theory of physics, the quantum theoretical developments have been concomitant with the quantum experimental findings. The requirement of the theoretical explanations of the observed emission spectrum of the blackbody radiation and the observations of the emission-absorption spectra from different materials led to the early twentieth-century (1900–1925) developments of quantum theory. Both these modern physics observations, being spectroscopic findings, refer to the observations of *intensity* distributions.

Although these observations led to the theories of the so-called 'modern' physics, making observations pertaining to intensity distributions was not the characteristic task of the modern (as opposed to 'old') experimental setups alone after all. The development of spectroscopic and interferometric experimental setups dates back to the seventeenth century – that is, much before the advent of quantum theory – when, for instance, Newton's experimental investigations in the domain of optics took place. The experimental setups in such investigations were designed to look for nothing but the intensity patterns (distributions) as the outcomes in different situations, such as dispersion, interference, diffraction, and so on. As an example of a (relatively) modern diffraction experiment, performed in the 1920s, we can think of the Davisson–Germer experiment that demonstrated the diffraction of electrons and the corresponding intensity distribution confirmed the de Broglie hypothesis.

Let us compare the descriptions of the classical and quantum polarization experiments involving light beams: in the former, the light beam is described in terms of classical electromagnetic waves, whereas in the latter, it is considered as made up of individual photons.

In the classical theory, an arbitrary polarization of the light beam is given by resolving its electric field vector (**E**) in the (time-dependent) 'horizontal' and 'vertical' components:

$$\mathbf{E} = E_{0x}\cos(\omega t)\,\hat{\mathbf{i}} + E_{0y}\cos(\omega t + \alpha)\,\hat{\mathbf{j}} = E_H(t)\,\hat{\mathbf{i}} + E_V(t)\,\hat{\mathbf{j}}, \tag{8.1}$$

with the squares of the classical electric field components giving the intensities of the two components.

In quantum theory, using the Dirac bra-ket notation, an arbitrary polarization of the light beam is given by resolving the quantum state of an individual photon ($|\Psi\rangle$) in the 'horizontal' and 'vertical' base states:

$$|\Psi(t)\rangle = c_H(t)|H\rangle + c_V(t)|V\rangle. \tag{8.2}$$

Similar to the 'classical' case, in the 'quantum' case one is interested in the observation of the distribution of the polarized intensities upon the passage of the light beam through a 'polarizing beam-splitter' system. With respect to the horizontal and the vertical base states, these intensities are given in terms of the probabilities, that is, the modulus-squares, $|c_H(t)|^2$ and $|c_V(t)|^2$, of the (so-called) probability amplitudes $c_H(t)$ and $c_V(t)$, see, for example, [9] for details. This way, the idea of 'intensity' can be seen to be connected to that of 'probability'.

Let us consider the nuclear magnetic resonance (i.e. NMR) setup, which is also called the NMR *spectrometer*. It represents an example of a relatively modern spectrometry[6]. It is used to probe the *intensity distribution* of nuclear spin states from a liquid state or a solid state sample. In NMR, the usual practice is to work with an ensemble of identically prepared identical systems (i.e. nuclear spins) for the determination of the expectation values – for which the usual term is 'ensemble average' – of single (say $\langle \hat{A} \rangle$) or correlated (say $\langle \hat{A} \cdot \hat{B} \rangle$) physical properties (i.e. those given by the 'joint' probability distributions) of micro-systems in different situations; see, for example, [10].

The intensity distributions in the polarization and NMR experiments are respectively of photons and nuclei, and are the functions of their 'polarization' and 'spin' states. In the experiments involving the LASER cooling of the dilute atomic gases – such as the setups employed for creating Bose–Einstein condensates – the intensity distribution is of the atoms of the gas and is a function of the 'velocity' states of those atoms (see, e.g., [11]). All these examples indicate that 'intensity distributions' are the primary outcomes of the quantum experimental setups.

Now, the 1920s' formulation of quantum physics (developed by Heisenberg, Born, Jordan, Dirac, Schrödinger etc.) represented a system of a 'universally valid' mechanics that offered new dynamical laws[7]. Also, this 'new' formulation had satisfactorily explained most of the experimental observations. And, the formal mathematical structure – which was developed by von Neumann as a mathematically rigorous re-expression of the quantum formulation in terms of the transformation theory and the theory of Hilbert spaces, and which resulted in the quantum-probabilistic formalism – came in 1932.

Considering this chronology, it is important to keep in mind that, although today the idea of 'quantum probability' can be described in purely mathematical terms by isolating it from the quantum experimental contexts, the quantum probabilistic formalism was not *already available* as a purely mathematical or statistical language for its use in the 1920s' developments of quantum mechanics. Rather, the development of an appropriate mathematical foundation – on which the quantum probabilistic formalism would rest – became possible *with hindsight*, that is, by following the observations of the already performed experiments that hinted at the

6 Still, the NMR technique is now more than eighty years old!

7 For instance, the 'statistical' interpretation of $|\Psi|^2$ had already been proposed by Max Born in 1926, see Subsection 8.1.3.

physical behaviour of micro-systems (e.g., in terms of intensity distributions in different situations) and, by following the previous (i.e. the 1920s') version of the quantum mechanical formulation.

In this regard, it is also important to note that, in the light of quantum physical experiments (in which intensity distributions are the observables), the notion of probability can make sense *only with reference to an ensemble* of similarly prepared systems. And, therefore, in Chapters 9 and 10 what we employ is the idea of an ensemble of *agents* in analogy with the idea of an ensemble of *electrons* in the double-slit interference context.

8.1.3 Quantum Theory: Foundational Problems and Interpretations

The probabilistic interpretation of quantum theory does lead to the explanation of quantum experimental observations, though the experimental observations themselves do not force such an interpretation. And, owing to this fact, the observations do not prevent the wave function from being understood in different ways [12]. Indeed, the quantum theoretical framework admits of a variety of interpretations (Born, Copenhagen, Pilot Wave, Many Worlds, Ensemble etc.) of the wave function. For a detailed discussion on different interpretations of quantum theory, see, for example, [7].

With reference to its intricate connection with the previous (classical) theories of physics, the issue of interpreting quantum theory has given rise to mainly two (and somewhat related) foundational problems. These are

1. The **measurement** problem: This problem refers, among other things, to the questions of whether 'wave function collapse' occurs during the measurement process and if yes, how. It is best expressed in terms of the notorious 'Schrödinger Cat' paradox. For a detailed discussion, see [13].
2. The problem of **quantum-to-classical** transition: This problem, also known as the problem of the appearance of classical world within quantum theory, is usually investigated in terms of the well-known decoherence programme [14]. For specific application of the programme, see, for example, [15].

It can be said that the different interpretations of quantum theory – in some way or another, and either directly or indirectly – aim to address the preceding problems. In the following we briefly explain some selected interpretations (see [16] for a discussion on the overlaps between and the mutual complementary nature of these interpretations):

1. **The Born Interpretation (BI):** Max Born was influenced by Einstein's remark related with the connection between the 'wave field' and the 'light quanta'

that "the waves only have to be wherever one needs to know the path of the corpuscular light quanta" and hence, by his idea of a "ghost field" [17]. Born wanted to establish a similar connection between 'electron waves' and 'corpuscular electrons' and therefore, to formulate the laws of electron motion in a manner analogous to the motion of light quanta (photons). He thus analyzed the quantum mechanical process of the 'electron–atom' collision by considering the incident electron as a plane wave and the scattered electron as a superposition of waves outgoing in different directions. Unlike the case of colliding macro-systems, for example, billiard balls, in which the unique 'scattered' (i.e., the 'outgoing') directions of motion can be predicted if the incident (i.e. the 'incoming') momentum directions are known, Born saw that for the case of a quantum collision process, such a predictability is eliminated on account of the wave description of the participating micro-systems. He explained the elimination of predictability in terms of the paradox that the motion of the scattered particle obeys the laws of probability, although the propagation of probability is in accord with causal laws. Born argued that this probability is proportional to the modulus-square of the wave function.

2. **The Copenhagen Interpretation (CI):** The CI is considered mainly as an outcome of the interactions between Niels Bohr, Werner Heisenberg, and Wolfgang Pauli, which took place during the 1920s and which were focussed on exploring the meaning of quantum theory. From the historical point of view, there does not exist any precise description of the interpretation, although the history of quantum mechanics suggests that one of the motivations behind coming up with the interpretation was the need to reconcile the idea of 'quantum jumps' with Schrödinger's wave mechanics. Also, some records show that there were certain disagreements between the views of Bohr and Heisenberg on how the theory could be interpreted (see e.g. [18], [19], [20]). Yet, some of the main features of the CI could be stated as follows:

   - According to CI, quantum mechanics is intrinsically indeterministic.
   - The wave function of the system obeys the Born rule (BI), according to which its modulus-square gives probabilities for the outcomes of measurements performed on the system.
   - Bohr promoted the principle of complementarity, according to which, systems possess certain pairs of properties (such as position and momentum) that cannot be determined simultaneously and hence, in this sense, those properties are complementary to each other.
   - Heisenberg insisted on a sharp distinction between the system, whose dynamics is to be observed and the apparatus (or the instrument) playing the role of an observer. According to him, the wave function represents available knowledge

of a system; and due to the change in available knowledge caused by the interaction of the system with an apparatus, it 'reduces' to a new state.

3. **The Pilot Wave Interpretation (PWI):** The PWI was originally proposed by Louis de Broglie and extended later by David Bohm to include measurements. According to the de Broglie–Bohm theory (DBBT), the evolution of the wave function is governed by the Schrödinger equation though there is not any process like wave function reduction. The DBBT is inherently nonlocal and is deterministic [12]. The PWI claims that particles always have definite positions, supports the notion of particle trajectory and considers the wave function as providing a guiding field for particles to move along definite space-time trajectories, with their motions governed by quantum potential energy derivable from the amplitude of the wave function.
4. **The Many-Worlds Interpretation (MWI):** The MWI promotes the idea of a universal wave function, whose evolution is understood as both deterministic and reversible at all times. It does not support the wave function collapse due to measurement [21]. Rather, it deals with the measurement process by applying the idea of decoherence, according to which correlations between the states of a system and the states (often called the 'pointer' states) of its environment are formed during their inevitable and entangling interaction. In other words, the part of the wave function describing the observer-states becomes increasingly entangled with the part of the wave function describing the observed-states. This way, each single observer–observed correlation state leads to the splitting of the universe into mutually unobservable though coexisting alternate histories.
5. **The Ensemble Interpretation (EI):** The EI takes, supports, and extends the Born statistical interpretation and claims that the wave function does not apply to an individual system – for example, a single particle – but is an abstract statistical quantity that only applies to an ensemble (a vast multitude) of similarly prepared systems or particles [22]. With respect to the contributions from Einstein, Born, and Ballentine in shaping up the EI, we prefer to call it the Einstein–Born–Ballentine interpretation.

Given this elaboration on the theoretical, experimental, and philosophical aspects of the physical foundations of quantum theory, it can be remarked that the quantum theory is much more than just a stand-alone, purely mathematical or statistical toolbox. And, ignoring these aspects and focussing only on its mathematical formalism would mean a quite one-sided use of quantum theory in the social sciences. Thus, by a 'quantum physical foundations focussed' approach to modelling we mean a way of describing and analyzing the processes and activities taking place in the world of economics and finance by developing (as far as possible) precise economics or financial analogues of the *physical* foundations – and not merely borrowing the mathematical language – of quantum theory. The economics/finance models constructed with such an approach will be called quantum physics–like (quantum physics–analogous).

8.2 Agent-Centric Quantum Physics-like Models: Expectations

In this section we set out specific analogy-driven expectations for quantum physics-analogous models – that is, models to be built that incorporate the quantum physical foundations discussed in the previous section. In doing so, we will have provided some hints of the exhaustiveness of such an approach.

The micro-origins of the happenings in the economics/finance world can be considered to lie in the individual agents' continuous arrival at specific choices by making decisions based on their preferences and on their subjective interpretations of the relevant socio-economic environments. Therefore, it will be useful to explain the expectations from the quantum physics–analogous models by giving a reference to some specific situation, for instance, the evolution of the preferences of agents during the 'trading' process. Here we give a qualitative description of such an evolution, which will turn out to be suitable for the purpose of constructing detailed quantum physics–analogous model (Chapters 9 and 10).

Consider a trading situation (e.g. buying or selling a financial instrument or a product at a particular price) that presents an individual agent with certain alternatives. The evolution of the agent state (represented in terms of the preferences) can be understood as a process having the following three stages:

- The *from* Stage: this stage marks the beginning of an agent's evolution in terms of her/his desire or intention to participate in a future transaction at some preferred price,
- The *via* Stage: at this stage the formation of a specific preference takes place in the agent's mind (possibly depending on their past experiences and cognitive biases),
- The *to* Stage: this stage marks the actualization of a potential choice the agent can end up making, subject to the influence of the relevant economics/financial condition on it.

Keeping in mind the preceding process as a prototype, in the following subsections we provide an outline of some of the expectations from the resulting quantum physics–analogous models in terms of a set of topics from (quantum) physics, thereby touching all the aspects of the quantum physical foundations discussed in the previous section.

8.2.1 Clarity of the 'System' in question and its environment

One of the centrally important notions for constructing a theoretical model – whether classical or quantum – of any physical phenomenon or process is that of a physical system, whose state is subject to dynamical evolution. Whether stated directly or via an indirect reference, physicists use different ways to categorize and label physical systems, such as a *simple* versus a *complex* system; a *single-particle* versus a *multi-particle* system; a *matter* system versus a *field* system; a *closed* versus an *open* system and so on.

A clarity of the category of the physical system in question thus becomes essential in the construction of the models of the reality. Also, realistic situations necessarily require modelling of the (local) environment for the system as well. As a simple example, we again consider the motion of a stone under gravity. In this situation the stone is modelled as a single-particle system and the Earth becomes its environment providing the gravitational field through which the motion takes place.

Thus, the foremost expectation from a quantum physics–analogous model for an economics or finance situation is to give a clear indication of *what is the system in question* and *what is its environment*. The process of the evolution of the preferences mentioned earlier clearly points to the *agent* as the system in question, for it is the 'preference' state of the agent-system that is undergoing the evolution. At the moment, the description seems to imply that it is a free evolution of the agent state, in the sense that the environment of the agent is not clearly specified, though we do consider and hence, model it in Chapter 10.

8.2.2 An Ensemble of Identical Systems

Once the system is identified in a given socio-economic problem, the quantum physics–analogous model is expected to clarify what it could mean by an *ensemble* of such *identically prepared identical systems*. The quantum physics–analogous economics/financial meanings of these terms are not only important for constructing quantum physics–analogous models, but are also essential for conducting the relevant simulations for the economics/finance domain.

We propose that two agent-systems are said to be identically prepared identical systems – in the physics–analogous experimental sense – if both of them are participating in the *same* economics or financial context, despite the possible behavioural heterogeneities. An *ensemble* of such agent-systems could then be understood as providing a set of differences in their behavioural response to the common context. With reference to the process described earlier, such differences can be understood in terms of the preferences of the agent-systems. The quantum physics–analogous model will then be expected not only to introduce an appropriate (i.e. as per the context) parameter quantifying the preferences but to also provide a physics-like interpretation for the same.

8.2.3 Stating the Dynamical Law

Given that the system in question is the agent-system and hence, that it is the arbitrary state of the agent-system that is supposed to undergo dynamical evolution, the next question is: which dynamical law should be assumed for such an evolution in our approach to the quantum physics–analogous modelling? The obvious answer is that the state evolution of the agent-system should be governed by a Schrödinger-like dynamical law. In other words, the agent-state evolution should be given by a Schrödinger-like equation. If this is the case, then the resulting model will be expected to provide analogous (i.e. physics-like) interpretations of the variables and the constants from the equation. For instance, in such a model there will have to be some centrally important function, analogous to the quantum wave function Ψ, to be used for describing the state of the agent-system. Moreover, it should be possible to give a position-like representation for such a function in terms of some appropriate independent variable capable of admitting a position-like interpretation.

8.2.4 Clarity of the Interpretational Stance

The above requirement leads to the next issue: that of the interpretation of Ψ in a quantum physics-analogous manner. With a Schrödinger-like equation used in the economics/finance domain, the resulting model will need clarifying the quantum interpretational stance and hence, will need developing a precise economics/financial analogue of Ψ accordingly.

For example, if a Born-like statistical interpretation of the wave function Ψ is chosen, then it will be expected from the resulting model that Ψ can be somehow identified with the total probability amplitude that the process (such as that given earlier) takes place. To be more precise, it should relate to the amplitude that the agent-system – when presented with certain alternatives within a given economics/financial context – *starts from* the desire for participating in that context and *arrives at* a choice corresponding to the context.

Such an identification will lead to a further expectation of clarifying how Ψ could obey the laws of combining amplitudes (in the different economics/finance analogues of quantum experiments) corresponding to those alternatives. The alternatives presented to the agent-system will, therefore, be expected to play the role of the economics/finance analogues of the quantum *base* (or *basis*) states. Following from the discussion of the previous subsection, it can also be said that, in general, as a legitimate quantum-like wave function, Ψ will be required to admit economically or financially meaningful representations with respect to different (e.g. position-like, momentum-like etc.) bases, so that, the square of its modulus, that is, $|\Psi|^2$ – with reference to the Born interpretation – will give the distribution of the probability density over different base-choices.

If, instead, a dynamical role of the wave function is assumed as primary – by following the de Broglie–Bohm pilot wave interpretation – and the statistical interpretation of the wave function is taken as secondary, then the resulting quantum physics–analogous model will be expected to provide a causally transparent and ontologically clear analysis of the activities of agent-systems from the world of economics and finance, apart from being able to predict the probability distributions in different contexts. In this case, the amplitude (R) of the wave function Ψ will be expected to give rise to the economics/financial analogue of the quantum potential energy ($Q \propto \nabla^2 R/R$).

In this case, the notion of the space-time trajectory of a particle – used in the classical physical theories of motion and supported by the pilot-wave interpretation, but generally disregarded in the conventional quantum interpretation – will be expected to play an analogous and ontologically meaningful role in the resulting model. Therefore, in the approach to the construction of – what could be called – the quantum physics–analogous *Bohmian* models, there will have to be an analogous economics/financial notion of the trajectories of the agent-systems from the ensemble, belonging to a suitably and analogously defined configuration space. Such agent-trajectories (for instance, with reference to the process described at the beginning of this section, evolving from the initial preferences to the final actualized choices) will be subject to some force-like quantity, F, acting on the agent-system during the entire process and having an expression $F = -\nabla(V + Q)$, in terms of the classical (V) and the quantum (Q) potential energy-like functions. The 'quantum physical foundations focussed' approach will then be expected to provide an economically or financially sensible interpretation of these potential energies.

It would be interesting to see what it could mean by the analogous economics/finance version of the measurement problem if the conventional interpretation of quantum theory is chosen to follow. In that case, it would be intriguing to have, for example, a quantum physics–analogous economics or finance description of the 'Schrödinger cat paradox.' The 'quantum physical foundations focussed' approach can be expected to lead to the analogous economics/financial version of the problem of quantum-to-classical transition. With respect to this, we think a truly successful quantum physics–analogous model will be that which would provide some room for accommodating the emergence of a *classical physics–analogous* behaviour of the economics/financial systems.

It is apparent that a quantum physics–analogous model – living up to the previous expectations – will make us more answerable, in terms of rigour and consistency and may allow a much wider scope for the investigation of the economics/financial activities, than the pure quantum math–analogous models. In other words, the 'quantum physical foundations focussed' approach will prove to be more exhaustive than the one that seems to have been applied to the development of the quantum math–analogous models.

With this background, it can be already anticipated that, the approach leading to quantum math–analogous models will appear relatively more straightforward than that leading to the quantum physics–analogous models, due to the intricacies arising from the requirement of taking into account the physical foundations in the latter, which seems to have been bypassed in the case of the former. Our discussions imply that, due to such a bypassing, the physical insight offered by pure quantum math–analogous models of the socio-economic reality may be limited.

We now proceed to discuss different criteria for evaluating analogical arguments and see how those criteria together with the quantum physical foundations focussed approach could guide us to the development of quantum physics–analogous models for the economic sciences. As indicated at the beginning of this chapter, the next two sections are devoted to a detailed explanation of this.

8.3 Analogies and Analogical Arguments: Selected Topics

Understanding the nature of analogies and framing the criteria in terms of which the analogical arguments could be evaluated are centuries-old quests for investigation. In the modern era, these quests come under the academic discipline of the philosophy of science. There is a huge literature devoted to the detailed discussions on the intricacies of these topics (see [23], [24] and the references therein). A detailed review covering all those intricacies shall obviously go beyond the scope of this book. However, given our ultimate task of constructing quantum physically *analogous* models, it is essential and useful to know – at least at an elementary level – the way in which the philosophers of science deal with these issues. For this, we follow the material presented in [24][8]. The survey presented in this section will be helpful also in setting the terminology for further discussions.

Therefore, in this section we offer glimpses of the corresponding investigations. We do it in terms of a survey of

- the various roles of analogy,
- the form of an analogical argument and,
- the theories that put forth different criteria for evaluating analogical arguments.

[8] The Stanford Encyclopedia of Philosophy (SEP) holds an exclusive license, granted by the author of the cited entry, to distribute its content in electronic and digital formats, and the editors have granted us permission to reproduce and distribute an extended excerpt from this entry in that format.

In simple words, an analogy is a comparison of similarities between two things (e.g. physical objects, concepts, notions, domains of knowledge). An analogical argument is a statement that points to the accepted (or well-known) similarities between those things to draw conclusion(s) about plausible further similarities. Mostly, the purpose of exploring analogy is to provide explanation or clarification via analogical reasoning.

8.3.1 Some Examples of Analogy

We begin with a few examples of analogy (from mathematics and physics), which will be useful in the rest of our discussion in this section:

1. **Rectangles and Boxes**: Rectangles (as 2D objects) and boxes (as 3D objects) have certain similarities. For example, the properties 'perimeter' and 'area' of rectangles can be considered as respectively analogous to the properties 'surface area' and 'volume' of boxes. Further, for a fixed value of the perimeter, by varying the dimensions of a rectangle, it can be found that, the rectangle occupying maximum area happens to have the same value of length and breadth, that is, it should be a square. Based on this finding, it can be conjectured that for a fixed value of surface area, the box occupying maximum volume should have the same value of length, breadth and height, that is, it should be a cube.
2. **Geometry and Algebra**: According to historians of science (see, e.g., [25]), the *form* (geometry) and the *number* (algebra) had been perceived as analogous by Pythagoras and Plato. Due to Descartes' analytical procedures, the correspondences between them became even more apparent. This way, it became possible to handle *geometric* problems using *algebraic* methods.
3. **Sound and Light**: There are certain similarities between sound and light, with reference to their behaviour (e.g. echo of sound is similar to reflection of light) and also, with reference to their properties (e.g. the 'sound' attributes of pitch and volume are similar respectively to the 'light' attributes of colour and brightness).
4. **Electric Current and Fluid**: Ohm's law speaks about the proportionality between the current (I) through a conductor and the potential difference (voltage, ΔV) across its end points. With the proportionality constant as the resistance (R) of the conductor, the corresponding relation is given by $\Delta V = RI$. Similarly, for an incompressible, Newtonian fluid in a pipe, its volumetric flow rate is found to be proportional to the pressure difference along the pipe. This is nothing but Poiseuille's law, given by the relation $\Delta P = K\frac{dV}{dt}$, with K as a constant. By recalling that the current can be understood as the charge flow rate (i.e. $I = \frac{dq}{dt}$), it can be seen that the mathematical expressions of Ohm's law and Poiseuille's law have an identical form.

5. **Continuity Equations**: In physics, the *motion* of a conserved entity (which can be massive or charged particles) and the *flow* of a conserved quantity (which can be energy or probability) are given by continuity equations. All the continuity equations display an identical mathematical (integral or differential) form. For example, the differential form of a continuity equation is given by

$$\frac{\partial \rho}{\partial t} + \nabla \cdot \mathbf{J} = 0. \tag{8.3}$$

   Here, ρ is the density, that is, the amount (of the entity or the quantity) per unit volume and $\mathbf{J}$ is the corresponding flux, that is, the amount moving or flowing per unit area per unit time.
6. **Quantum Dynamics and Classical Hydrodynamics**: Following the Madelung approach, the equations of quantum dynamics can be expressed in terms of the velocity and the probability density fields. These equations and the equations of classical hydrodynamics have an identical mathematical form [12].

8.3.2 The Roles of Analogy

Specifically with reference to the natural sciences, the philosophers of science identify two different (though, apparently interdependent) roles of analogy ([24]). These are:

1. **Heuristic:** If an analogy aids discoveries by giving rise to insights that can further lead to the formulation of possible solutions to problems then its role is said to be heuristic.
2. **Justificatory:** In the case when an analogical argument can be used to lead to some conclusion, it is said to have played the justificatory role.

Yet another role of analogy – called the **Programmatic** role – is an inseparable blend of both the preceding roles, namely of discovery and justification ([26]). Programmatic analogies are those that can lead to the development of new directions within a given programme of research or even lead to an altogether new programme. Such analogies can help guide conceptual development and in some cases, drive the theoretical unification of two different domains (see the second example from Subsection 8.3.1.).

8.3.3 On the Form of an Analogical Argument

An analogical argument is often expressed by referring to two domains or the fields of knowledge: S (the *source* domain) and T (the *target* domain). These domains are, respectively, physics and the economic sciences in this book. Here, a 'domain' can be understood as a collection of elements under different categories such as objects, properties, relations, and functions. It also includes a set of interpreted and accepted statements about those objects, properties, relations, and functions. Therefore, exploring analogies between the domains S and T involves an investigation of similarities, or more precisely, one-to-one mapping between different elements (i.e. objects, properties, relations, and functions) belonging to the S domain and those belonging to the T domain.

Here we would like to add that sometimes it may be the case that certain similarities between S and T are known even before getting into a proper reasoning, which would involve the elements from the former based on their analogies with the corresponding elements from the latter. Alternatively, it may be the case that similarities between two domains may not be obvious to begin with, but can be made apparent by adopting a certain viewpoint, thereby making them justifiable on the basis of plausibility arguments to motivate further investigation. In either of these cases, it is useful to have a certain statement, which could qualify as an analogical argument for systematic further exploration of analogies between the two domains.

Now, in general, premise(s) and conclusion(s) are essential ingredients of any argument, and analogical arguments are no exception. In terms of these ingredients, the generic form of an analogical argument can be expressed in the following manner ([24]):

There are certain known similarities between S and T. (*Premise*)
The domain S has some further feature Q. (*Premise*)
$\Longrightarrow$ T also has the feature Q, or some feature Q^* similar to Q. (*Conclusion*)

In logic and the philosophy of science, such an argument form is called *ampliative*, which means 'extending' or "adding to that which is already known," though there is no guarantee that the conclusion will follow from the premises. Analogical arguments stated in this way certainly do have relevance to our 'physics – economic sciences' analogy exploration and serve as a guide to our direction of further investigation in this book (see Section 8.4).

8.3.4 Components of Analogy

Just like having a well-defined form for expressing analogical arguments, as we will see in our upcoming discussions, it is useful to have a certain terminology to classify analogies. With reference to [27], the components of an analogy can be described as follows:

1. **Positive Analogy:** Let $P \equiv \{P_1, P_2, \cdots, P_n\}$ be the set of accepted propositions about the source domain S. Suppose that $P^* \equiv \{P_1^*, P_2^*, \cdots, P_n^*\}$ is the corresponding set of accepted propositions about the target domain T, such that P

and P^* represent the accepted (or known) similarities. Then P is referred to as the *positive analogy*.

2. **Negative Analogy:** Let $A \equiv \{A_1, A_2, \cdots, A_r\}$ be the set of accepted propositions about the source domain S. Also, let $B^* \equiv \{B_1^*, B_2^*, \cdots, B_s^*\}$ be the set of accepted propositions about the target domain T. Suppose that the analogous sets of accepted propositions, $A^* \equiv \{A_1^*, A_2^*, \cdots, A_r^*\}$ and $B \equiv \{B_1, B_2, \cdots, B_s\}$, respectively about T and S, fail to hold. Then A and B are referred to as the *negative analogy*.
3. **Neutral Analogy:** Let N be the set of accepted propositions about the source domain S, such that it is not known whether there exists an analogous set of accepted propositions in T. Then N is referred to as the *neutral analogy*.
4. **Hypothetical Analogy:** If one of the elements (i.e. an accepted proposition about the source domain S), say Q, from the set N of the neutral analogy is used to conjecture that an analogous proposition, say Q^*, is true in T, then the conjecture can be referred to as a hypothetical analogy.

The preceding components can be represented in a compact, tabulated manner as shown in Table 8.1.

As we will see, the premise of the analogical argument given in Section 8.4 is a plausibility expression leading to the identification of positive analogy and its conclusion is a hypothetical analogy. Here, the missing link is the plausibility of the hypothetical analogy itself – the question is: what makes a hypothetical analogy plausible? A hypothetical analogy is plausible when some reason can be provided, based on an epistemic support, to believe it and hence, to investigate it further even prior to testing.

With this background, we now turn to the discussion on some selected different theories of analogical arguments and reasoning.

Table 8.1 *Components of analogy between the 'source' (S) and the 'target' (T) domains.*

S	T	Component of Analogy
P	P^*	Positive
A	$\overline{A^*}$	Negative
$\overline{B}$	B^*	Negative
N	??	Neutral
$Q \in N$	Q^*	Hypothetical

8.3.5 Aristotle's Theory

The Aristotelian theory provides a model for evaluating analogical arguments that may appeal to common sense. It lays a foundation for all the later, more sophisticated and refined theories of analogical reasoning. It can be summarized in terms of the following four criteria for the evaluation of analogical arguments:

1. The strength of an analogy depends upon the number of similarities.
2. Similarity reduces to identical properties and relations.
3. Good analogies derive from underlying common causes or general laws.
4. A good analogical argument need not presuppose acquaintance with the underlying universal (generalization).

As can be seen from the first two criteria, the identification of similarities between the source and the target domains is of central importance in the Aristotelian theory. With reference to the previous subsection, we can say that these criteria imply that the *horizontal* relationships (i.e. those pertaining to the positive analogy, $P \leftrightarrow P^*$) provide guidance for the analogical reasoning. Also, the greater the overall similarity in the properties and relations of the two domains, the stronger the analogical argument.

8.3.6 Material Analogy vs. Formal Analogy

Modern theories of analogical reasoning agree with the Aristotelian theory with regards to the importance of the overall similarity between the source and the target domains. However, these theories tend to give different weights to the aspects of similarity, such as *material* and *formal*. For theories asserting the primacy of the material analogy (Subsection 8.3.7), similarity in the *observable* features of the two domains is the foremost requirement. Theories focussing on the formal analogy (Subsection 8.3.8) require instead a high-level *structural* correspondence. In the following, we briefly explain the meaning of these kinds of analogy.

1. *Material Analogy.* An analogy between the domains S and T can be understood as *material*, if the domains have *observable* or *pre-theoretic* similarities ([28]). The similarity between the phenomenon of echo (in case of sound) and that of reflection (in case of light) is an example of an observable similarity in the two domains. Also, such a similarity is called pre-theoretic in the sense that, it can be noticed without knowing or without making any reference to the theories of sound and light that provide explanations of these phenomena. In other words, a material analogy speaks about similarities rooted in the *factual knowledge* of the source and target domains ([29]).
2. *Formal Analogy.* An analogy between the domains S and T can be understood as *formal*, mainly under the following two situations:

- a property from S and that from T share abstract similarities; for example, the perceptual property of *pitch* of sound and the perceptual property of *colour* of light.
- a relation from S and that from T have an identical mathematical form; for example, the continuity equations in physics (see 8.3.1).

In light of this distinction, the development of the quantum math–analogous models seems to be assuming a *formal* analogy between the quantum formalism and the social sciences, in the sense of the mathematical expressions from these two domains having identical mathematical form. At this point we would like to remark that having identical mathematical form does not guarantee that the quantities involved in the mathematical expressions corresponding to the quantum math–analogous models possess analogous (quantum *physics*–like) interpretations in the corresponding economics/finance contexts.

8.3.7 Hesse's Theory

Hesse's theory can be regarded as a modified version of the Aristotelian theory. In [28], she puts forth the following three criteria for evaluating the acceptability of an analogical argument:

1. *Requirement of Material Analogy* – The horizontal relations ($P \leftrightarrow P^*$) must include similarities between observable properties.
2. *Causal Condition* – The vertical relations (i.e. the relations within the source domain) must be causal relations in some acceptable scientific sense.
3. *No-Essential-Difference Condition* – The essential properties and causal relations of the source domain must not have been shown to be part of the negative analogy.

The discussion from the previous subsection indicates that in the case of the theories of sound and light, we have both material and formal analogies. However, Hesse's theory seems to imply that the material analogy is independent of the formal analogy and that the formal analogy is not the starting point but rather the end result of analogical reasoning. For example, the material similarity, which can be interpreted as the similarity between an aspect of the behaviour of sound and that of light motivates the investigation of some more, plausible correspondences, such as those between pitch and colour, or between volume and brightness. Thus, the condition of material analogy can be seen as the foremost requirement in Hesse's theory.

The criterion of causal condition calls for the existence of a causal connection between the positive analogy and the hypothetical analogy, with the latter expressed in terms of a plausibility argument for having an analogous feature in the target domain. In Hesse's words ([28]), "the essential requirement for a good argument from analogy is a tendency to co-occurrence, i.e., a causal relationship: 'The vertical relations in the model (source) are causal relations in some acceptable scientific sense, where there are no compelling a priori reasons for denying that causal relations of the same kind may hold between terms of the explanandum (target).'" The criterion of no-essential-difference condition requires definition of essential, or at least some clarity on what it could mean by essential. Hesse's theory suggests that a property or relation can be considered as essential, if there exists a close, causal connection of the property or relation with the known positive analogy.

Philosophers of science point out certain weaknesses in Hesse's criteria ([24]), mainly with reference to the examples of analogy from mathematics. Indeed, in mathematics, a material analogy – as defined by Hesse – seems to have no role to play and deductions follow from formal analogies. Also, the problem with the *no-essential-difference* condition is that it can (unnecessarily) rule out the possibility of a good analogical argument that starts with already accepted or acceptable analogies. Though there are many essential differences between rectangles and boxes, there are certain similarities too, which can be exploited to provide good analogical arguments. Without having reference to a specific inferential context, it becomes impossible to identify relevant differences, decide their essentiality and weigh those differences against the similarities. Of course, in spite of such weaknesses, Hesse's theory represents a notable progress in our grasp of analogical reasoning.

8.3.8 The Structure-Mapping Theory

Structuralist theories are those which apply a model-theoretic isomorphism to identify similarities between S and T and to provide analogy-based reasoning. The structure-mapping theory (SMT) has been the most influential in this category. It is proposed by Gentner ([30]) and assesses the quality of analogies in terms of their structural aspects. According to the SMT, analogies are about the similarities in the relationships between objects of S and those of T. SMT asserts that the content of an analogy is determined by the interrelationships between the facts (i.e. the structural properties) rather than simple isolated features. The core idea of SMT is the *principle of systematicity*, which is stated as: "A predicate that belongs to a mappable system of mutually interconnecting relationships is more likely to be imported into the target than is an isolated predicate." ([30])

To clarify this, the SMT distinguishes between two basic kinds of predicate:

a. propositions asserting properties of the individual objects, e.g. "an object E is small in size."

b. propositions asserting relationships between two or more objects, e.g. "the object E revolves around the object S."

This way, a higher-order predicate asserts relationships between propositions. Thus, according to the SMT, a systematic analogy must involve correspondence (i.e. mapping) between high-order relations and their components. It seems to imply that 'systematicity' of the associated analogy mapping dictates 'plausibility' of conclusions that follow the mapping.

However, according to the critics of the SMT ([24]):

- It is neither necessary nor sufficient to have greater systematicity for ensuring a more plausible analogical inference. In other words, an increased systematicity will not guarantee increased plausibility.
- It is possible to have a high degree of structural parallelism that is exhibited by an implausible analogy.
- The systematicity requirement seems to ignore causal relevance, whether positive or negative. Consequently, it may even weaken the inference in causal analogies.
- With reference to the distinction between an object-rich domain and a relation-rich domain ([31])[9] it has been remarked that the SMT works better in the case of investigating analogies between relation-rich domains (e.g. the classical mechanics model of the solar system and the analogous Rutherford–Bohr quantum model of an atom).
- The reliability of the systematicity principle is poor in arriving at plausible analogical inference, when the considered pair of domains happens to be object-rich.

8.3.9 Bartha's Articulation Model

With respect to the approaches discussed in Subsections 8.3.5, 8.3.7 and 8.3.8, it can be said that the main focus of the Aristotelian approach is enumerating similarities, Hesse's theory proposes an identification of observable, pre-theoretic correspondences and Gentner's theory insists on seeking mappings involving complex, high-level networks of relations. However, in some cases neither of these approaches is relevant/useful. For instance, the analogical argument for the similarities between rectangles and boxes (example 1, 8.3.1) can be evaluated neither by counting similarities nor by looking for observable resemblances nor by seeking structural correspondences between the two kinds of geometric notions.

For such cases, Bartha proposes a different approach, in terms of the following two conditions [23]:

[9] An object-rich domain is the one in which the number of objects is greater than the number of relations (and properties). Otherwise the domain can be understood as relation-rich.

- *Prior Association* – There must be a clear connection, in the source domain, between the known similarities (the positive analogy) and the further similarity that is projected to hold in the target domain (the hypothetical analogy). This relationship determines which features of the source are critical to the analogical inference.
- *Potential for Generalization* – There must be reason to think that the same kind of connection could obtain in the target domain. There must be no critical disanalogy between the domains.

This approach suggests beginning with an articulation of the prior association of the elements from the source domain. Recalling example 1, (8.3.1), it is the association between the notion of a rectangle, its attributes (e.g. perimeter, area) and the specific proofs for the results that are generally true for all the rectangles. Then, Bartha's approach asks for identifying the critical features of the source (i.e. 'rectangles') domain. With reference to a given proof, these are the concepts and assumptions used in the proof. Upon identification of the critical features from the source domain, the approach requires an assessment of the potential for generalization. This assessment is to be done by investigating whether there are analogues of those critical features in the target (i.e. 'boxes') domain ([24]).

8.4 Analogies and Analogical Arguments: Key Takeaways

The previous section provided an overview of several perspectives on the identification of analogies between two domains and on the evaluation of analogical arguments. In the light of these perspectives, the present section develops a strategy for applying quantum mathematical language and the physical foundations of quantum theory to construct generic quantum physics–like models for economic-financial activities.

The key takeaways can be summarized as follows:

1. *The Source and Target Domains.* Analogies and analogical arguments refer to the objects, properties, relations and functions belonging to two domains of knowledge; one of them is to be identified as the source ($\mathbb{S}$) and the other as the target ($\mathbb{T}$).
2. *Types of Analogies.* At the interface of $\mathbb{S}$ and $\mathbb{T}$ four distinct types of analogies can be identified: positive, negative, neutral and hypothetical.
3. *Hesse's Theory* – one may start with the identification of the pre-theoretic similarities in the *behaviour* of the objects from $\mathbb{S}$ and $\mathbb{T}$ under analogous conditions and then, based on such an identification, hypothesize further similarities.
4. *Gentner's Structure-Mapping Theory* – one may explore how the *relations between the attributes* of the objects from $\mathbb{T}$ can be introduced in analogy with

those from $\mathbb{S}$ and investigate how those relations may lead to a model-theoretic structural isomorphism between $\mathbb{S}$ and $\mathbb{T}$.

5. *Bartha's Articulation Model* – one may begin by considering the known similarities between the elements (i.e. objects / properties / relations / functions) of $\mathbb{S}$, say $\{S_1, S_2\}$ and those from $\mathbb{T}$, say $\{T_1, T_2\}$, so that, the element T_1 is similar to S_1 and T_2 to S_2. Next, one can take into account the connection between the $\mathbb{S}$-domain elements from those similarities (i.e. S_1 and S_2) and some further element from $\mathbb{S}$, say S_3. Subject to the potential for generalization, one can then hypothesize that an analogous connection exists between the 'known similarity' elements from $\mathbb{T}$ (i.e. T_1 and T_2) and some further element, say T_3.

The last three points – corresponding to the three theories of analogical arguments – from the above list indicate that the problem of framing criteria for evaluating analogical arguments is an ongoing topic of research. Keeping in mind this fact, in the following we consider the preceding summary as a guideline for our undertaking, viz. systematic development of analogies between (quantum) physics and the economic sciences, though, we avoid following only one of the preceding theories, for we do find the core philosophies of *all* these proposals appealing.

As we will see, a customized combination of all these theories along with the terminologies introduced in the previous section turn out to be useful in our exploration. The purpose of the present section is to elaborate on how the philosophical standpoints of each of the theories (summarized in the last three points in the preceding list) could be woven together to form the basis for approaching the task of developing quantum physics–analogous models for economics and finance.

8.4.1 The Four-Step Procedure

Applying the very first takeaway from the preceding list, we now formally declare that, hence onwards, throughout our exploration, physics (and, in particular, quantum physics) will play the role of the source domain, which we denote by $\mathbb{S}_P$ here, whereas economics and finance will together play the role of the target domain (denoted by $\mathbb{T}_{EF}$ here). And, taking into consideration all the takeaways together, we propose the following four-step procedure for going about developing quantum physics–analogous models for the $\mathbb{T}_{EF}$ domain:

1. **Step 1:** *Identify* pre-theoretic qualitative similarities pertaining to the behaviour of the objects from $\mathbb{S}_P$ and those from $\mathbb{T}_{EF}$.
2. **Step 2:** *Hypothesize* further similarities in the $\mathbb{T}_{EF}$ domain by (a) noting the relations between the object-attributes and the role played by those relations in the construction of the dynamical models from the $\mathbb{S}_P$ domain, and (b) assuming that, it is possible to construct models for the $\mathbb{T}_{EF}$ domain bearing structural similarities with those from the $\mathbb{S}_P$ domain.

3. **Step 3:** *Select* a suitable quantum physical model that is, a model belonging to the $\mathbb{S}_P$ domain. With reference to the model, *introduce* new physics-like notions useful for constructing a corresponding quantum physics-analogous model, i.e. a model belonging to the $\mathbb{T}_{EF}$ domain and *indicate* the $\mathbb{S}_P \leftrightarrow \mathbb{T}_{EF}$ mapping.
4. **Step 4:** Use the $\mathbb{S}_P \leftrightarrow \mathbb{T}_{EF}$ mapping to *construct* a quantum physics-analogous model for (at least some of the aspects of) the reality corresponding to the $\mathbb{T}_{EF}$ domain. *Provide* physics-like interpretations of the parameters (quantities) from the constructed quantum physics–analogous model.

The similarities to be identified by implementing **Step 1** will collectively form the *positive* analogy (See Subsection 8.4.2). **Step 2** will take us to the *hypothetical* analogy at the interface of $\mathbb{S}_P$ and $\mathbb{T}_{EF}$ (See Subsection 8.4.3). This way, the philosophies discussed in Subsections 8.3.7 (i.e. Hesse's "material analogy" criterion), 8.3.8 (i.e. the 'systematicity' principle of the structure-mapping theory) and 8.3.9 (i.e. the 'prior association' criterion of Bartha's articulation model) will be partially followed, albeit in a customized manner. **Step 3** and **Step 4** are implemented in Chapters 9 and 10 respectively.

8.4.2 The "Positive" Analogy

We already anticipate that the analogy exploration at the '$\mathbb{S}_P \leftrightarrow \mathbb{T}_{EF}$' interface could be way more intricate than that at the source–target interfaces corresponding to the examples mentioned in Subsection 8.3.1. With reference to our discussions from Sections 8.1 and 8.2, such an exploration may be treated as a special case that could potentially bring new perspectives to the field of economics and finance and perhaps, even to that of the philosophy of analogical arguments and reasoning. In a sense, this situation also justifies our preference of not strictly adhering to the philosophy of any one particular analogical argument theory.

Also, one can be skeptical even on the existence and hence, on the possibility of identifying pre-theoretic similarities between physics and the economic sciences. This could be due to the commonsensical (and up to a certain extent, seemingly correct) understanding that these domains are entirely different bodies of knowledge, apparently without any connection or any overlap whatsoever between their respective aims and also, between their respective modelling and problem-solving approaches.

Despite this skepticism, as we will see, the activities of the world of economics and finance can be seen as analogous to the happenings of the physical world[10].

[10] Recalling the spirit in which we presented some of the glimpses into the history of physics (Section 1.1), here we are totally ignoring the fact that the physical world, by its very definition, includes any and every *happening*, including human activities, which do take place *within* the universe and this way, are very much a part of the physical world.

Already in Section 8.2 we have seen how the "agent system" from an economics or finance context could play a role analogous to the system in physics. Thus, the key is to view the happenings of the economics/finance world as originating from the desires, preferences and decisions of the 'agent systems' (or simply, agents). Here, an agent system can be understood as either an individual (person) or a group (or a collection) of individuals participating in a given economics or finance context (for example, buying or selling of a financial instrument or a product).

The entire process of behavioural decision-making (for example, starting from the intention to sell or buy something to actually ending up selling or buying at some given price) by the agent-systems can be understood as analogous to the evolution (change) of states of physical systems. Moreover, in the economics/finance world it is routinely observed that different social situations influence the preferences and decisions of the agent systems in different ways. We claim that this is similar to the way different physical environments (forces / fields / potentials) influence the evolution of the states (position states / momentum states / energy states) of physical systems in different ways.

In physics, the notions of 'system,' 'environment' and 'state evolution' – whether used implicitly or stated explicitly – are vital and their clear specification is essential in the construction of the dynamical models of reality. Given this fact, it can be said that the 'agent-centric' view of the T_{EF} domain – which tends to support the notions analogous to system, environment and state evolution – makes to some extent apparent those similarities between the domains $\mathbb{S}_P$ and $\mathbb{T}_{EF}$ that are absolutely essential for developing physics-analogous dynamical models of the economics/financial activities.

Following Hesse's terminology, we call these the 'pre-theoretic' similarities and also, as stated earlier, interpret them as the elements of the *positive* analogy at the interface of the two domains (see Table 8.2).

This completes **Step 1**. We now turn to **Step 2**.

Table 8.2 *Pre-theoretic similarities between the source and target domains, interpreted as the elements of positive analogy.*

Category	Element $\in \mathbb{S}_P$	Element $\in \mathbb{T}_{EF}$
Object (System)	physical systems	agent systems
Object (Surroundings)	forces/fields/ potentials	socio-economic environments
Dynamics (Evolution)	influence on: position/energy states	influence on: preferences/decisions

8.4.3 The 'Hypothetical' Analogy

As seen previously, in physics, the evolution of the states of a system subject to the influence of an external force or potential is given by the solutions to the equation(s) of motion representing dynamical laws. In the corresponding problems, often the "system" in question is precisely identified and is distinguished from its "environment". In principle, the environment is the rest of the universe, though, often, in actual model calculations, only the most relevant part of the rest of the universe (i.e. the local environment) is considered. This "most relevant" part leads to the situation-specific expression of the force (F) or the potential energy (V), which is given in terms of the physical attributes of *both* the system and its environment. For example, in the 'motion under gravity' situation, the expression for the force is $F = mg$, where, m (mass) is the system attribute and $g = \frac{GM}{r^2}$ (i.e. the acceleration due to gravity) is the environment attribute. Similarly, in the 'motion of a charged particle in electrostatic field' situation, the expression for the force is $F = qE$, where q (charge) is the system-attribute and E (electrostatic field) is the environment-attribute. We invite the reader to verify this point by considering a few more situations and the corresponding expressions of the forces or the potential energies.

Thus, considering the table from the previous subsection and given that there exists a pre-theoretic one-to-one correspondence between the elements from the $\mathbb{S}_P$ and the $\mathbb{T}_{EF}$ domains under the categories of object, surroundings and dynamics, we **hypothesize** two further similarities at the interface of the two domains:

- **Similarity 1 – Identical Mathematical Form:** The agent-state evolution – under the influence of the agent's socio-economic environment for the contexts from the $\mathbb{T}_{EF}$ domain – can be expressed through equations having mathematical forms identical to some of the dynamical equations (e.g. the equation for Newton's second law, the Hamilton–Jacobi equation and the Schrödinger equation) from the $\mathbb{S}_P$ domain.
- **Similarity 2 – Analogous Interpretation:** The quantities (to be defined), which would be involved in those equations (i.e. from the $\mathbb{T}_{EF}$ domain) can be given interpretations that will be analogous to their physics counterparts belonging to the $\mathbb{S}_P$ domain.

Now, again, one may argue that the *identical* mathematical form for the 'source' and the 'target' domain equations automatically ensures an *analogous* interpretation of the corresponding quantities that make up the two equations – so that, there is no need of stating the hypothetical 'Similarity 2' separately. For instance, consider the fourth example (Electric Current and Fluid) from Subsection 8.3.1. The relations following from Ohm's law and Poiseuille's law are $\Delta V = R\frac{dq}{dt}$ and $\Delta P = K\frac{dV}{dt}$. Not only do these relations have identical mathematical form, but it can be argued that the corresponding quantities also have analogous physical interpretations. Same goes for the fifth example, that is, of the continuity equations.

A possible reason for this could be that these are the examples of processes from the sub-domains of physics, whereas, as we have already remarked, our endeavour involves two, apparently quite distinct, unrelated domains: physics (the source domain) and the economic sciences (the target domain). Therefore, even if we carry over (merely) the mathematical forms of the dynamical equations from the $\mathbb{S}_P$ domain to the $\mathbb{T}_{EF}$ domain and somehow stick labels – for the quantities involved in those forms – that may appear relevant to economics or finance, there is no guarantee that the usage of the corresponding mathematical equations in the $\mathbb{T}_{EF}$ domain will automatically ensure physics-analogous interpretations of those quantities.

In fact, even *within* the domain of physics there are several examples where, for the quantities involved in analogous mathematical relations, the physical interpretations are *not at all* analogous. To explain this point, we consider the following pair of mathematical expressions:

$$\mathbf{L} = I\,\boldsymbol{\omega} \tag{8.4a}$$

$$\mathbf{P} = \alpha\,\mathbf{E}. \tag{8.4b}$$

These expressions have identical form in the pure mathematical sense. Indeed, in three dimensions, the angular momentum **L**, the angular velocity $\boldsymbol{\omega}$, the polarization **P** and the electric field **E** are all 3-vectors, which could be given 3×1 column matrix representations and, both the multiples, the moment of inertia I and the polarizability α are, in fact, tensors with '3×3' square matrix representations w.r.t. the corresponding bases. For instance, for the expression '$\mathbf{L} = I\,\boldsymbol{\omega}$' we have:

$$\begin{pmatrix} L_x \\ L_y \\ L_z \end{pmatrix} = \begin{pmatrix} I_{xx} & I_{xy} & I_{xz} \\ I_{yx} & I_{yy} & I_{yz} \\ I_{zx} & I_{zy} & I_{zz} \end{pmatrix} \begin{pmatrix} \omega_x \\ \omega_y \\ \omega_z \end{pmatrix} \tag{8.5}$$

with mathematically similar expression for the 'polarization-electric field' relation. However, neither the quantities appearing in the two LHS (i.e. **L** and **P**) nor the vectors and tensors from the corresponding RHS (i.e. the pairs $\boldsymbol{\omega}$, **E** and I, α, respectively) have analogous physical interpretations.

This implies that, the mathematical language itself – in spite of the precision and definiteness it provides – is blind to the *physical* interpretation of the quantities involved in an equation. And therefore, physics as the source domain cannot be taken merely as the source (or provider) of mathematical relations and methods. We emphasize that developing analogies between physics as the source domain and any other as the target domain is much more than just borrowing the mathematical forms of the relations from physics.

8.4.4 Quantum Math–like vs. Quantum Physics–like

The above discussion completes **Step 2** toward the construction of quantum physics-analogous models for the $\mathbb{T}_{EF}$ domain by drawing analogies from the $\mathbb{S}_P$ domain, whereas the implementations of **Step 3** and **Step 4** are the topics of the next two chapters, respectively.

Before proceeding to these steps we would like to reiterate the distinction between the two ways of developing quantum models for economics and finance: the first – and as noted in the last three chapters, the more established – is the one making use of the mathematical language of quantum theory in economics and finance. The second one – as we insist here – proposes to pay a serious attention to also its interpretations and physical foundations. The first way gives (predominantly) the quantum math-like models, whereas the second one is expected to lead us to the quantum physics-like models. The latter can be understood as a generalization of the former.

References

[1] Baaquie, B. (2018). *Quantum Field Theory for Economics and Finance*. Cambridge University Press: Cambridge, UK.

[2] Feynman, R., Leighton, R. and Sands, M. 1965. *The Feynman Lectures on Physics, volume 2*. Basic Books.

[3] Schwinger, J. (1989). A Path to Quantum Electrodynamics. *Physics Today*, 42 (2), 42–48.

[4] Iemhoff, R. (2020). Intuitionism in the Philosophy of Mathematics. *The Stanford Encyclopedia of Philosophy* (Fall 2020 Edition), Edward N. Zalta (ed.), https://plato.stanford.edu/archives/fall2020/entries/intuitionism/.

[5] von Neumann, J. (1955). *Mathematical Foundations of Quantum Mechanics*. Princeton University Press.

[6] Moretti, V. (2017). *Spectral Theory and Quantum Mechanics*. Springer.

[7] Norsen, T. 2017. *Foundations of Quantum Mechanics: An Exploration of the Physical Meaning of Quantum Theory*. Springer: Cham.

[8] Athalye, V. and Haven, E. (2021). Socio-Economic sciences: beyond quantum math-like formalisms. *Quantum Reports*, 3, 656.

[9] Beck, M. (2012). *Quantum Mechanics: Theory and Experiment*. Oxford University Press: New York, NY, USA.

[10] Athalye V. S., Roy S. S. and Mahesh, T. S. (2011). Investigation of the Leggett-Garg inequality for precessing nuclear spins. *Physical Review Letters*, 107, 130402.

[11] Anderson, M., Ensher, J., Matthews, M., Wieman, C. and Cornell, E. (1995). Observation of Bose–Einstein condensation in a dilute atomic vapor. *Science*, 269, 198–201.

[12] Holland, P. (1993). *The Quantum Theory of Motion*. Cambridge University Press: Cambridge.

[13] Dürr, D. and Lazarovici, D. (2020). *Understanding Quantum Mechanics: The World According to Modern Quantum Foundations*. Springer: Cham.

[14] Schlosshauer, M. (2007). *Decoherence and the Quantum to Classical Transition.* Springer: Berlin.
[15] Athalye, V. and Kumar A. (2006). Decoherence of superposition of molecular chiral states due to Rayleigh scattering. *Journal of Physics B: Atomic, Molecular and Optical Physics*, **39**, 2633.
[16] Athalye, V. (2025). Unifying Perspectives: All-encompassing Reality vs. All-inclusive Physics. *Journal of Physics: Conference Series.*
[17] Born M. (1927). Quantum mechanics of collision processes (Original in German: Quantenmechanik der Stoβvorgänge). *Zeit. Phys.* **38**, 803.
[18] Omnès, R. (1999). *Understanding Quantum Mechanics.* Princeton University Press: Princeton, NJ.
[19] Faye, J. (2019). Copenhagen Interpretation of Quantum Mechanics. *The Stanford Encyclopedia of Philosophy.* (Winter 2019 Edition), Edward N. Zalta (ed.), https://plato.stanford.edu/archives/win2019/entries/qm-copenhagen/.
[20] Baggott, J. (2011). *The Quantum Story.* Oxford Landmark Science.
[21] Wallace, D. (2012). *The Emergent Multiverse: Quantum Theory According to the Everett Interpretation.* Oxford University Press: Oxford.
[22] Ballentine, L. (1970). Statistical interpretation of quantum mechanics. *Reviews of Modern Physics*, 42, 358.
[23] Bartha, P. (2010). *By Parallel Reasoning: The Construction and Evaluation of Analogical Arguments.* Oxford University Press: New York.
[24] Bartha, P. (2022). Analogy and Analogical Reasoning. *The Stanford Encyclopedia of Philosophy.* (Summer 2022 Edition), Edward N. Zalta (ed.), https://plato .stanford.edu/archives/ sum2022/entries/reasoning-analogy/.
[25] Singer, C. (1941). *A Short History of Science to the Nineteenth Century.* Oxford Clarendon Press.
[26] Kuhn, T. (1996). *The Structure of Scientific Revolutions.* 3rd Edition. University of Chicago Press: Chicago.
[27] Keynes, J. (1921). *A Treatise on Probability.* Macmillan: London.
[28] Hesse, M. (1964). Analogy and confirmation theory. *Philosophy of Science*, 31, 319.
[29] Norton, J. (2010). There are no universal rules for induction. *Philosophy of Science*, 77, 765.
[30] Gentner, D. (1983). Structure-Mapping: A theoretical framework for analogy. *Cognitive Science*, 7, 155.
[31] Schlimm, D. (2010). Two ways of analogy: Extending the study of analogies to mathematical domains. *Philosophy of Science*, 75, 178.

9

Single-Particle Systems versus Agents – I

This chapter (together with Chapter 10) introduces what is probably the highlight of the book, that is, it attempts to answer the important question: what can we now do with the quantum physics–like stance? An immediate, almost obvious, discussion centres around the analogies with the famed double-slit experiment. We set ourselves the task of answering how we can begin to enumerate, quite precisely, analogies between electrons and agents. As the reader will find out, we will need to move over several (important) hurdles, one of them being the perennially difficult analogy we need to make with the Planck constant. We then proceed in shaping the idea of two-preference interference, a concept of paramount importance in our quest to properly define the quantum physics–like research direction.

As noted in the preceding chapter, there are definite ways in which analogies between the elements (i.e. objects, properties, relations, and functions) belonging to two domains can be identified and categorized. There, several alternative approaches proposed for evaluating analogical arguments have been discussed. And, by combining certain selected philosophical elements, taken from all those approaches, a pathway – in terms of a four-step procedure – was proposed to the construction of quantum physics–like models for economics and finance via identifying positive analogies, hypothesizing further analogies and introducing new physics-like notions.

Following the positive analogies, which were considered as the pre-theoretic similarities, it was hypothesized that not just the *mathematical form* – identical to that for some of the central dynamical equations from physics – but also, the precise *physics-like interpretations* of the newly introduced quantities should work in an analogous manner in the economics-finance domain. It is now clear that such a hypothesis would essentially require an application of the physical foundations of quantum theory, if a particular economic/financial model is to qualify as a truly quantum physics–analogous model. It was also anticipated that the more we pay critical attention to such details, the more we become answerable. So we explained how – being more exhaustive – the 'quantum physical foundations focussed' approach could be more demanding than compared to the approach which considers the construction of pure quantum math–analogous models. The corresponding enhanced answerability was explained in terms of the expectations from such an approach and therefore, from the resulting quantum physics–analogous models.

Thus, it is time to see how this new approach actually works when it comes to the development of useful models for the economics/finance domain. In other words, it is time to implement **Step 3** and **Step 4** from the procedure outlined in Subsection 8.4.1. **Step 3** is concerned with the selection of a model from the physics domain and the introduction of new physics-like notions useful in a certain (economics/finance) context. This step should also indicate the mapping between the two domains with reference to the context. Now, there exists a huge variety of (quantum) physics models, with different models appropriate and useful in different contexts.

Physicists classify the models of reality in several ways, for example, by indicating:

- the *paradigm-* or *domain*-type to which the models belong, for example, Newtonian models, non-relativistic models, relativistic models, quantum models, relativistic quantum models, quantum field models and so on.
- the *system-type* for which the models are applicable, for example, models for single-particle systems, models for many-body systems, models for two-state systems, models for multi-state systems and so on.
- the *problem-type* or *condition-type*, for example, models for scattering versus models for the bound-state problems, models for the time-independent problems versus models for the time-dependent problems, models for the equilibrium condition versus models for the non-equilibrium condition and so on.

Anticipating its potential usefulness in the agent-centric view of the economic/financial activities, we pick up the quantum model for the behaviour (motion) of single-particle systems. And, with reference to the pedagogical structure of [1], we can be even more specific: for simplicity, we can consider single-particle systems with *two* base-states. We can begin with the implementation of **Step 3** by recalling one of the most familiar problems under this category, namely, the double-slit interference. As it will turn out, the quantum double-slit model helps provide an agent-centric physics-analogous model for the microdynamics underlying the seemingly stochastic, macrolevel evolution of the stock share prices.

The purpose of this and the next chapter is to present, in a detailed and comprehensive manner, the double-slit interference-like quantum model for the behaviour of individual agents (market participants) as the elements of an ensemble. In doing so, we will also elaborate on how the kinematic and dynamic quantities for describing and analyzing the agent behaviour in the economics/finance domain could be introduced in analogy with the kinematic and dynamic quantities defined in the physics domain for describing and analyzing physical systems' behaviour. We begin by providing an outline of the double-slit interference phenomenon when the single-particle systems are 'electrons'.

9.1 Quantum Behaviour of Electrons: Two-Slit Interference

In 1801, Thomas Young proposed the wave theory of light, in contrast to the then century-old view due to Newton, that light possessed a particle (corpuscular) nature. The demonstration that Young developed to support his theory was that of the 'interference' of light, which is now a well-established phenomenon in physical optics. Based on the principle of superposition of waves, the analysis of the resulting interference pattern – consisting of maxima and minima of intensity – required the modelling of light as a wave.

Later on, in 1905, when Einstein provided an explanation of the observation of the photoelectric effect, following Planck's hypothesis, he had to invoke the corpuscular nature of light. Today, in the scientific community it is generally accepted that light possesses a *dual* (i.e. wave–particle) nature. Considering the case of light, whose nature was proposed first as particle-like and then understood as wave-like and then again as particle-like – depending on the experimental context, in the 1920s physicists began wondering whether such a dual nature works for matter as well. It was Louis de Broglie (1924) who, for the first time, hypothesized a wave nature for matter (e.g. electrons), with such (so-called) *matter waves* representing a spatial distribution of phase (see [2] for details). This hypothesis was confirmed by the famous Davisson–Germer experiment, which demonstrated electron diffraction.

To point out the seemingly mysterious aspects of the *quantum* behaviour of physical systems, in the very first chapter of [1], Feynman considers an electron two-slit interference experiment. For this, he analyzes first the behaviour of bullets and that of waves in similar situations and then, compares and contrasts the behaviour of electrons with each of them. It can be said that, since then, the electron two-slit interference situation is being used as an important topic in the account of quantum mechanics and of its foundations, not only to convey the apparently mysterious *quantum* behaviour of physical systems but also to explain and possibly justify one's interpretational stance, see, for example, [3].

In the following we explain briefly the electron two-slit interference experiment. We will not get into the intricacies of an actual experiment [4]. Rather, we will focus only on those facts that are directly relevant to the further discussions.

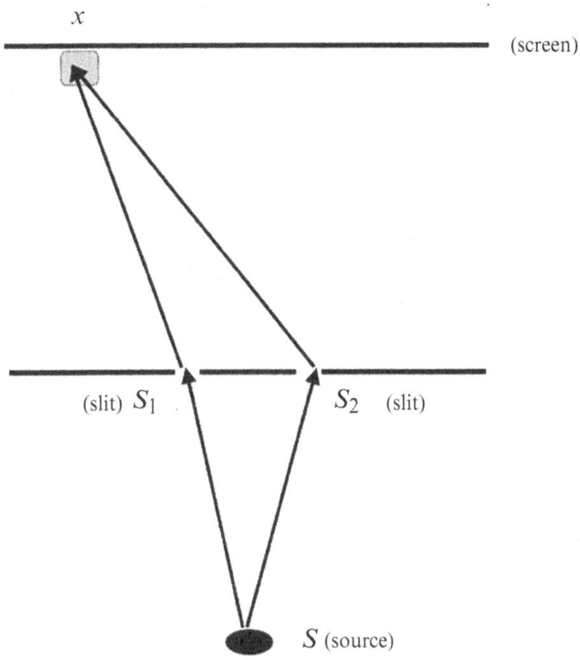

Figure 9.1 Electron two-slit interference.

As shown in Figure 9.1, the experimental setup consists of a source of electrons (s), a pair of slits (s_1 and s_2), and a screen (with a movable detector placed at an arbitrary position x). The experiment involves observation of the distribution of intensity of electrons upon their arrival at the screen after starting from the source and passing through the slits. The electron source can be theoretically modelled as an ensemble of identically prepared identical systems (i.e. electrons), and hence, each electron coming out of the source can be understood as an element from such an ensemble. The journey of each of the electrons from the source to the screen can then be considered as a trial and with a huge number of electrons starting from the source and arriving at the screen, we have a situation involving a large number of repetitions. This way the observed "intensity distribution" of the electrons on the screen can be given a theoretical explanation in terms of the "probability distribution" as the function of electron position, that is, $P(x)$, on the screen.

In terms of the Dirac bra-ket notation $\langle\cdot|\cdot\rangle$ the probability amplitude that an electron starts from s (the *starting* condition) and gets detected at a position x on the screen (the *final* condition) can be expressed as: $\langle x|s\rangle$. The experimental setup for the electron two-slit interference provides two paths for the electron to reach the detector: one via slit s_1 and the other via slit s_2. If slit s_1 is kept open and s_2 is kept closed, the electron starting from s may arrive at the position x by passing through s_1. The probability amplitude for this possibility, say $\phi_1(x)$, can be expressed symbolically as

$$\phi_1(x) = \langle x|s_1\rangle\langle s_1|s\rangle. \tag{9.1}$$

The corresponding probability is given by

$$P_1(x) = |\langle x|s_1\rangle\langle s_1|s\rangle|^2 = |\phi_1(x)|^2. \tag{9.2}$$

Similarly, if slit s_2 is kept open and s_1 is kept closed, the electron starting from s may arrive at the position x by passing through s_2. The probability amplitude for this possibility, say $\phi_2(x)$, can be expressed symbolically as

$$\phi_2(x) = \langle x|s_2\rangle\langle s_2|s\rangle. \tag{9.3}$$

The probability in this case is given by

$$P_2(x) = |\langle x|s_2\rangle\langle s_2|s\rangle|^2 = |\phi_2(x)|^2. \tag{9.4}$$

However, if both the slits are kept open, it turns out that the probability $P_{12}(x)$ of the arrival of an electron at x is not simply the sum of the two probabilities given by Equations 9.2 and 9.4, that is,

$$P_{12}(x) \neq P_1(x) + P_2(x), \qquad \textbf{(electrons)} \tag{9.5}$$

rather, it equals the absolute square of the sum of the amplitudes corresponding to the two possibilities, that is,

$$P_{12}(x) = |\phi_1(x) + \phi_2(x)|^2 = |\phi_1(x)|^2 + |\phi_2(x)|^2 + 2\,\mathrm{Re}[\phi_1^*(x)\phi_2(x)]. \tag{9.6}$$

The third term on the right-hand side of Equation 9.6 is called the interference term. This term, due to the undetectably small de Broglie wavelengths, would be negligible for bullets, if an electron two-slit interference-like experiment were performed for such macrosystems (see section 1-6 of [1]). Therefore, for macrosystems the usual additive relation for the total probability works well, that is,

$$P_{12}(x) = P_1(x) + P_2(x). \qquad \textbf{(bullets)} \tag{9.7}$$

Considering the preceding description of the electron two-slit interference (and its subsequent dynamical analysis in Chapter 10), we can say that, if we aim for the construction of an analogous quantum model for an economics or finance context, then we will have to introduce quantities analogous to the kinematic and dynamic physical quantities, such as position, velocity, momentum, force, kinetic and potential energies. We will also have to introduce space-time-like arena (i.e. some notion analogous to space-time) for the economics-finance world. This is done in the next section.

9.2 From Electrons to Agents: The Analogy Checklist

In Section 8.2 we provided a qualitative description for the evolution of agent preferences with respect to a typical trading scenario. There we briefly discussed the expectations from the 'quantum physical foundations focussed' approach and hence, from a resulting quantum physics–analogous model. In doing so, we mentioned some of the central concepts from quantum physics (such as the idea of an ensemble of identically prepared, identical systems and the interpretation of the wave function) and also, advanced issues from quantum foundations (such as the measurement problem and the problem of quantum-to-classical transition), for which the analogues should be clearly stated or developed for the economics/finance domain. In Section 8.3, starting from Aristotle's theory, we discussed different criteria for good analogical arguments in terms of the modern approaches to theorizing analogical reasoning: Hesse's Theory, Gentner's Structure-Mapping Theory and Bartha's Articulation Model. There we also distinguished between different kinds of analogy, such as positive and negative; neutral and hypothetical; and, material and formal.

Now, with the background of the electron two-slit interference context, we can proceed in a more definitive manner, that is, with the aim of the development of an analogous model for the economics/finance domain. With several hints already given in an abstract manner (Sections 1.2, 8.1, 8.2, 8.4) for constructing quantum physics–analogous models, the purpose of the present section is to introduce and define some physics-analogous notions and ideas for the economics/finance domain, which will be used in Section 9.3 and Chapter 10 for developing and analyzing a model for agent dynamics in analogy with the electron two-slit interference model. Since the electron two-slit interference fairly exemplifies almost all the essential aspects of quantum theory by serving as a genuine, prototype 'quantum' model, the present section can also be understood as providing an analogy checklist useful for developing general quantum physics–analogous economics/finance models.

9.2.1 Space-time–like Arena for the Economics World

In Section 1.2, a viewpoint on the generic structure of the physics frameworks was presented. As discussed there, with hindsight, we may regard a theoretical physics framework as a combination of a certain "dynamical paradigm" and a "spacetime structure" appropriate to the paradigm[1]. By 'appropriate' we mean that the dynamical laws of the paradigm remain invariant under the coordinate transformations on the space-time.

[1] In physics, a space-time structure is interpreted as a background (an arena or a stage) and, in general, depending on the physics paradigm (say, Newtonian physics, special relativity, quantum mechanics, quantum field theory etc.), it could be interpreted as having a *fixed* structure or a *dynamic* structure, or, even an *emergent* structure.

It may be argued that the space-time structure is an integral part of any dynamical paradigm of physics, in the sense that, once you specify the latter, the former gets specified as well, so that there is no need to state and consider the former as a different feature. However, we prefer to consider these features as distinct, though inseparable. For example, the framework of nonrelativistic mechanics is the combination of the Newtonian paradigm and the Galilean structure of the space-time. It is well known that the Newtonian laws of motion are invariant under Galilean transformation. Now, the Galilean structure of space-time is appropriate for expressing the (Galilean-relativistic) quantum physics paradigm *as well.* Indeed, the Schrödinger equation turns out to be invariant under Galilean transformation (see e.g. [5]).

To be able to describe agent-centric dynamical processes belonging to the world of economics and finance in a physics-like manner, we can define the space of agent 'preferences' $\mathbb{P}$ as the arena (with a structure that is isomorphic to the real space $\mathbb{R}$) where those processes can be assumed to take place. Also, in analogy with $\mathbb{R}^n$ we can assume a Cartesian n-dimensional representation for $\mathbb{P}^n$. This way each point $P \in \mathbb{P}^n$ can be associated with a unique set of Cartesian (real-numbered) position-like coordinates (i.e., preferences) $\mathbf{x} \equiv (x_1, x_2, \cdots, x_n) \in \mathbb{R}^n$.

The preference of an individual agent, in a given economics or finance context, can then be quantified by suitably defining the preference variable $\mathbf{x}$ analogous to the position variable in physics. Just like there is a unit (metre) for 'distance' (or length or position), depending upon its type, one may suitably define a certain unit for the preference under consideration. For instance, if a preference refers to 'price' (involved in buying or selling something), then, the unit of the preference may be taken as the currency itself (say, pounds or dollars). Such a preference unit would then enable the definitions of velocity-like and acceleration-like kinematic quantities defined on the space of preferences.

9.2.2 Meanings of 'Agent-System' and 'Economics-Finance' Context

As mentioned in the previous chapter, in the theories of physics, physical systems are classified and modelled in different ways: a *simple* or a *complex* system; a *single-particle* or a *multi-particle* system; a *matter* system or a *field* system and so on. Clearly, given our aim of constructing an electron two-slit interference-like model for the economics/finance domain, the idea of 'single-particle system' will be sufficient for developing useful analogies.

Modelling a physical system as a single-particle system amounts to ignoring its internal structure. Even its size does not matter. For example, in the analysis of planetary motion – that is, for the so-called central force problem – individual planets and the Sun are modelled as single-particle systems. This way of modelling is not very unrealistic, in spite of the fact that the overall motion-related pure mechanical considerations tend to ignore the complex internal structure of such huge objects, for such details have no role to play at the expected level of analysis of their motion.

Here, we interpret the notion of an agent-system in a manner analogous to the notion of a physical system. And, by the term 'agent-system' (or simply, agent) we mean a system of a single or many conscious human beings (e.g. individual citizens and taxpayers of a given country, participants in a financial market, a business firm or a company, the central bank of a nation etc.). Such an agent-system may become aware of and hence, be willing to participate in a certain economic/financial context. By an economic/financial context for the agent-systems we mean any circumstance or a situation happening within the world of economics and finance that would allow the agent-systems' participation through a well-defined role.

Recalling the example from Section 8.2, an economics or a financial context can also be understood as presenting several ways (alternatives) to an agent-system for participating in it. These will be later on called the base-preferences (in analogy with the idea of base-states or basis-states). An agent-system's willingness to participate in the context can then be understood as leading to the emergence of a particular *preference* – that is, preferring one alternative over the other – out of the agent-system's mind for the context. For example, an agent-system may have a willingness to participate in the 'financial market' context, by playing the role of a buyer or of a seller, or to participate in the 'oil market' context, by being in the demand camp or in the supply camp.

9.2.3 The Mass-like 'Social Inertia'

In physics, 'mass' of a system refers to the content of its matter and it is the measure of the inertia of the system. It also represents the strength (or lack thereof) of the system for being able to influence the motion of other physical systems. For example, we can compare the Earth with an electron. Both have well-defined masses, though with a huge difference in the values. The former, through its strong gravity, influences the motion of all the physical systems in its vicinity, whereas the (gravitational) influence of the latter on the other systems is not at all significant.

We can introduce the idea of 'social mass' of an agent-system and understand it as the measure of its opposition to any change in its state of rest or of uniform motion (within the preference space-time). In other words, the social mass of an agent-system can be interpreted as indicating how easily a change can be brought into the uniform dynamics of an agent-system and can also be understood as a measure of a given agent-system's influence on the motion of the other agent-systems.

For instance, in the context of a particular economic policy taken by a nation's central bank affecting the financial decisions of the nation's citizens, the central bank, as an agent-system, will play a role analogous to a (mass-wise) heavier physical system (such as the earth) and the individual citizens of the nation will play a role similar to (again, mass-wise) lighter physical systems (such as a stone or a ball gravitating to the earth).

Also, for 'social mass' as an inertia-like attribute of the agents, we may define a unit to be used to quantify its strength. For instance, it could be measured in terms of the *social mass unit* (s.m.u.), analogous to the *atomic mass unit*, (a.m.u.) with the corresponding quantity (so much s.m.u.) being proportional to the agent-system's net worth.

Unlike the case of the masses of physical systems, which show a great variety in their numerical values, though they follow the same measurement standards, we need not get into the intricacies of defining a 'common-to-all-contexts measurement standard' for the social masses of all the possible (macro/micro) agent-systems throughout the entire world. Tentatively, we may just choose to stick to an arbitrary scale specific to a given context. In a given economics or financial context, the social masses of the participating agent-systems, given their net worths, may range from 1 s.m.u. (e.g., for an individual citizen) to say, 100 s.m.u. (which could be a firm). Also, it is not necessary that the mass of 10 s.m.u. correspond to 10 individual human minds together. It may just be for a single individual having a net worth that is way more than the other agent-systems participating in the same context.

Thus, we do not assume that the value of mass in the unit of s.m.u. is numerically equal to the number of human minds. For example, there are economic/financial contexts, for which the corresponding dynamics directly involve business magnates or government organizations and corporations. These are always very powerful and influential and hence, in those contexts, can be modelled as agent-systems with very high social masses, despite being run by only a small number of human authorities that tend to control the activities of those organizations, in a manner that can affect the financial fortunes of (social mass–wise lighter) individual human agents.

9.2.4 Physics-like Dynamical Quantities for the Economics-Finance World

In terms of social mass (m), preference ($\mathbf{x}$) and background time (t) as the fundamental quantities, and also, by using velocity-like and acceleration-like kinematic quantities, we can obtain the derived, physics-like, dynamical quantities such as the force acting on individual agents and the momentum, the kinetic energy and the potential energy of an individual agent. Moreover, corresponding to conservative-type social forces (F), the notion of social (classical) potential energy (V) must be meaningful so that, the former, precisely in a physics-like manner, should be interpretable as the negative derivative (or more generally, gradient) of the former: $F = -\frac{dV}{dx}$.

9.2.5 *Meaning of 'Quantum' Processes in the Economics World*

Usually quantum-level processes involve micro-sized systems moving with high speeds and occurring over very short time scales, whereas the processes of the classical world, which emerge from those underlying quantum processes, involve macro-sized systems moving with relatively low speeds and take relatively longer durations to occur. Analogously, the economic/financial processes can also be categorized as quantum-like and classical-like, with reference to the size and the speed of the agent-systems involved in the process in question and also, with reference to the related dynamical time scales.

9.2.6 *Fixing the Value of the Planck-like Constant*

In quantum theory, Planck's constant has been interpreted as the 'quantum of action' (see e.g. [2]) and its SI unit is the product of the energy unit and time, i.e. joule · second. Given that, the energy unit (joule) is equivalent to the product of the units of mass (kg) and of velocity squared (m^2/s^2), the value of the Planck-like constant with reference to a given economic or financial context, for instance, may be set as the product of the orders of the magnitudes of three quantities: the *elementary social mass*, the *squared agent-velocity* for that context and the *time scale* representing the relevant dynamics of the context.

9.3 Quantum Behaviour of Agents: Two-Preference Interference

In this section, we propose to *map* the electron two-slit interference model (belonging to the physics domain and representing a set of elements, such as single-particle systems (i.e. electrons), source, slits, base states, screen, probability amplitude),[2] with a situation belonging to the economics/finance domain, involving the behaviour of agents in a typical trading scenario from the financial markets. With reference to the process described in Section 8.2, we call it the 'agent two-preference interference' situation. Following Section 9.1, we come up with a symbolic expression for the probability amplitude that *an agent starts from the fixed previous stock share price and arrives at a potential future price.*

Ultimately we will be concerned with a quantum physics–analogous model enabling the prediction of the probability distribution over future stock share prices

2 The reader may recall the discussion from Subsection 8.3.3.

in terms of the behaviour of the market participants (i.e. 'agents') taking place *between* two successive time-instants, at which definite values of stock prices can be observed on the corresponding time series. Such a time series is simply a sequence of prices realized at different instants as shown in (9.8):

$$\{p(t_1), p(t_2), p(t_3), \cdots, p(t_n), \cdots\} \tag{9.8}$$

For simplicity, we take the preceding sequence as a reference to visualize the financial market dynamics as split at its interface (i.e. at the formation of *definite* stock prices) in the macro and the micro subdomains of the world of economics and financial activities. Such a delineation is similar to the fuzzy classical–quantum (macro–micro) boundary for the physical world, which is usually understood in terms of the *size* parameter (or masses) for physical systems as the delineator and hence, is defined vaguely by very large-sized molecules. Recalling our brief remark from Subsection 9.2.5, we choose to refer to the dynamical *time scale* – over which different activities take place in the economics/finance world – as the parameter representing the required delineator. Thus, the two subdomains for the economics/finance world are as follows:

- **The Micro-domain:** From the preceding price-sequence, we notice that the price of the stock, which was settled as $p(t_1)$ at the instant t_1, changes to some other value, namely, $p(t_2)$, at the instant t_2. Thus, by micro-domain dynamics we mean the *underlying* dynamics *that could cause* such a change. It takes place over a relatively shorter time scale (on the order of seconds or even shorter) and refers to what goes on (i.e. $p(t_k) \longrightarrow p(t_{k+1})$) between the successive time instants t_k and t_{k+1}.
- **The Macro-domain:** Given the 'formed' stock prices, the macro-domain dynamics is the *overlying* subsequent dynamics *that could be caused* by the price changes (among other things). It can include the far-reaching, long-term effects on the local and global economies. The macro-domain thus involves the longer time scale dynamics, that could take place over a time scale ranging from days to weeks to even months.

Of course, such an oversimplified distinction between the two subdomains does not imply that these two kinds of dynamics can be treated and understood separately. In fact, they are far from being separable. As time goes on, they both serve as feedbacks for each other and keep influencing each other in a complex, elusive manner. Specifically, with regard to the micro-domain – which is our concern here – whatever happens between the successive instants t_k and t_{k+1} is the complex reality and coming up with a 'realistic' dynamical model for it is, no doubt, extremely challenging.

Nevertheless, as we have seen in the previous chapters, there have been numerous attempts to model the various aspects of the economic/financial reality. What we are about to present is also one such model of the reality. And, we admit that, like other models, this, being a model – and not the reality itself – it will be way too less complicated than actual reality.

Let us begin with an explanation of the analogues of the source, the slit, the screen and the probability amplitude, relevant to the agent two-preference interference model, by referring to the electron two-slit interference model.

9.3.1 The Source

For the electron two-slit interference model, the source is that of the electrons, whereas for the agent two-preference interference model, it is that of agents. Such an agent source, similar to the case of the electrons, can be theoretically modelled as an ensemble of identically prepared identical agent systems – namely a collection of agents participating in the same financial context with the individual agents having an intention to transact, that is, to buy or to sell stock shares at different prices. The agent source can be assumed to come to exist right after the formation of a particular stock share price (say, $p(t_k)$). And, the evolution of the arbitrary state of each of these agents – starting from the source – can be given a representation in terms of the agent preferences. This way, similar to the journey of each of the electrons in the two-slit interference situation, the journey of each of the agents can be considered as a single trial. With a huge number of agents having their individual preferences evolving, that is, taking them to the different potential future stock share prices (one of which to be actualized at the instant t_{k+1}), we have a situation analogous to the electron two-slit interference involving a large number of repetitions.

9.3.2 The Slit

For a huge collection of the agents having an intention to transact, each of those agents can be understood as falling in one of the following two categories:

A. a subcollection of agents, who intend to trade stock shares at a price that is *lower* than the previous price; and,
B. a subcollection of agents, who intend to trade stock shares at a price that is *higher* than the previous price.

Note that, by the subcollection under category 'A.' we do not necessarily mean buyers, although it can be argued that such a subcollection would correspond most likely to the buyers. The same goes for the subcollection of the agents under category 'B.' – the corresponding agents need not be sellers only. This way, we can allow a more realistic situation, in which there could be individual agents who are willing to sell their shares at a price lower than the previous and there could be some other agents who are willing to buy shares at a price higher than the previous.

In analogy with the electron two-*slit* interference situation, for our agent two-*preference* interference case, the preceding categories enable definitions of the orthogonal base-states, which we denote by $|-\rangle$ (corresponding to category 'A.') and $|+\rangle$ (corresponding to category 'B.'). These base states represent two mutually exclusive and exhaustive intentions of the agents and provide alternatives for each individual agent through which its state can evolve to definite (potential final) prices, starting from the agent source. In this manner, the states $|-\rangle$ and $|+\rangle$ play roles similar to the slits from the electron two-slit interference situation.

9.3.3 The Screen

In the electron two-slit interference situation the screen represents a set of different potential *locations* (different values of x) at which the electrons arrive and form an intensity pattern, which can be explained quantum theoretically in terms of a probability density distribution, following from Equation (9.6). Analogously, in the agent two-preference interference situation we can talk about a probability density distribution for the agent-systems' arrival at a 'conceptual' screen representing a set of different potential future *prices* for the instant t_{k+1}.

9.3.4 The Probability Amplitude

We now introduce the formal idea of the probability amplitude that an agent arrives at a potential future stock share price $p(t_{k+1})$, starting from a previous price $p(t_k)$, in a manner discussed in Section 9.1.

If the 'trade' dynamics were to allow only those trades for which the quotes correspond to the stock share prices *lower* than the previous price, then the probability amplitude for this possibility, say ϕ_-, can be expressed symbolically as

$$\phi_- = \langle p(t_{k+1})|-\rangle\langle-|p(t_k)\rangle. \tag{9.9}$$

The corresponding probability is given by

$$P_- = |\langle p(t_{k+1})|-\rangle\langle-|p(t_k)\rangle|^2 = |\phi_-|^2. \tag{9.10}$$

Instead, if the 'trade' dynamics were to allow only those trades for which the quotes correspond to the stock share prices *higher* than the previous price, then the probability amplitude for this possibility, say ϕ_+, can be expressed symbolically as

$$\phi_+ = \langle p(t_{k+1})|+\rangle\langle+|p(t_k)\rangle. \tag{9.11}$$

The corresponding probability is given by

$$P_+ = |\langle p(t_{k+1})|+\rangle\langle+|p(t_k)\rangle|^2 = |\phi_+|^2. \tag{9.12}$$

When the trades of both the categories are allowed, in analogy with the electron two-slit interference, each agent will be presented with two *alternative trading channels* by the market. Through these channels individual agents could begin with an initial preference for a price at which they would like to trade. Assuming an electron-like 'quantum' behaviour of each individual agent, the probability $P_\pm$ that an agent arrives at some future price $p(t_{k+1})$ should not simply be the sum of the two probabilities given by Equations 9.10 and 9.12, that is,

$$P_\pm \neq P_- + P_+, \qquad \textbf{(agents)}, \tag{9.13}$$

rather, it should equal the absolute square of the sum of the amplitudes corresponding to the two possibilities, that is,

$$P_\pm(x) = |\phi_- + \phi_+|^2 = |\phi_-|^2 + |\phi_+|^2 + 2\,\mathrm{Re}[\phi_-^*\phi_+]. \tag{9.14}$$

Similar to the case of the electrons, the third term on the right-hand side of Equation 9.14 can be considered as the term arising out of the interference between the agent-amplitudes corresponding to the two alternatives for trading. This way, the present section proposes a mapping between the electron two-slit interference (from the 'physics' domain) and the agent two-preference interference (from the 'economics/finance' domain). And, in this way the implementation of **Step 3** finishes.

References

[1] Feynman, R. P., Leighton, R. B. and Sands, M. (1965). *The Feynman Lectures on Physics, volume 3*. New York, NY: Basic Books.
[2] Baggott, J. (2011). *The Quantum Story*. Oxford: Oxford Landmark Science.
[3] Dürr, D. and Lazarovici, D. (2020). *Understanding Quantum Mechanics: The World According to Modern Quantum Foundations*. Cham: Springer.
[4] Tonomura, A. et al. (1989). Demonstration of single-electron buildup of an interference pattern. *American Journal of Physics*, 57, 117.
[5] Auletta, G., Fortunato, M. and Parisi, G. (2009). *Quantum Mechanics*. Cambridge University Press: Cambridge, UK.

10

Single-Particle Systems versus Agents – II

This chapter provides, we believe, for the apogee of what we think will form the base for success of the quantum physics–like applications. Readers are invited in this chapter to carefully study the two-slit interference experiment with electrons and, analogously, the agent two-preference interference, for a variety of classical potential energy functions. We provide for a discussion on free particles and a variety of linear potentials. Probability distributions and trajectories are studied for the array of said potentials. We hope this chapter can be inspirational to all those readers who also believe that the quantum physics–like research orientation is really the next important step to take.

In the previous chapter we described the electron two-slit interference experiment. There the probability amplitude and the corresponding probability – that an electron starts from the source and reaches at a point on the screen – were expressed symbolically in terms of the Dirac bra-ket (i.e. $\langle\cdot|\cdot\rangle$) notation (See Equation (9.5)). Also, an outline of how the 'agent two-preference interference' situation could be mapped with the 'electron two-slit interference' situation was provided.

Similar to the case of an electron, a symbolic expression for the total probability amplitude (and hence, for the probability) – that an agent arrives at a potential future stock share price $p(t_{k+1})$, starting from a previous price $p(t_k)$ – was given. Now it is time to implement **Step 4** from the procedure discussed in Section 8.4. It has to do with the construction of a quantum physics–analogous *dynamical* model and also, with the interpretation of the model parameters. This is done in the present chapter.

For this, we begin with a brief analysis of the quantum dynamics of the two-slit interference phenomenon in terms of the evolution of the electron wave function, $\psi(x,t)$ and imitate it (in an analogous manner) for the agent two-preference interference situation. Just like electrons, we consider the agents as single-particle systems and with reference to the goal mentioned in Section 9.3, focus on the agent dynamics taking place between two successive instants, t_k and t_{k+1}, respectively corresponding to an already set previous price, $p(t_k)$ and the next potential price, $p(t_{k+1})$, of a stock share. The resulting agent two-preference interference model provides a way to predict the probability density distribution over a range of the potential (future) prices, similar to the case for the electrons, which enables the prediction of the probability density distribution over a range of the potential electron positions on the screen.

For the case of electrons, following Section 2.5 and following the Einstein–Born–Ballentine statistical interpretation of quantum mechanics (Subsection 8.1.3), we take the quantum state given by ψ as representing an ensemble of identically prepared electrons and the function $|\psi|^2 = R^2$ gives the probability density[1] of *finding an electron* within an infinitesimal volume element around different potential positions on the screen. Analogously, for the case of agents, ψ would represent an ensemble of identically prepared agents and $|\psi|^2$ would then be understood as the probability density of *finding an agent* within an infinitesimal price element around different future prices.

The phrase 'probability density of finding a particle' indirectly refers to an observer 'looking for the particle' and hence, to the act of measurement. Such an interpretation of the wave function tends to regard quantum theoretical models as concerned purely with the *epistemic* (i.e. knowable or verifiable) aspects of natural processes. In other words, such a viewpoint considers the processes of *state preparation* and *state detection*[2] as the only processes that matter in the construction of the quantum model for the situation in question. It can be said that, for the case of electrons, their journey from the *source* to the *slits* conceptualizes the former process, whereas the *screen* conceptualizes the latter one. Consequently, the only goal set for the quantum theoretical analysis is to derive the evolution of the wave function from the initial state to the final state and hence, to obtain the distribution of probability density (R^2), to account for the intensity distribution as observed on the screen.

Of course, the analysis of a quantum (experimental) situation and hence, the construction of the corresponding quantum model, need not be based purely on epistemic grounds. For instance, for the case of electrons, we can ask: could the actual motion of the individual electrons – as the elements of the ensemble – be conceptualized in terms of the trajectories connecting their 'individual' initial positions (at the slits) with their 'individual' final positions (at the screen)? Could each of the trajectories be understood as a trial, so that the *distribution of the trajectory*

[1] In three dimensions, $|\psi|^2$ is the probability per unit volume, whereas in one dimension it is the probability per unit length.

[2] Or, as Asher Peres calls them, *preparations* and *tests*; see [1].

density – arising out of a huge number of such trials – approximates the *distribution of probability density* and hence explains the intensity distribution as observed on the screen?

Answering such questions requires what could be called an *ontological* perspective, that is, a perspective that

(a) makes one wonder what might possibly be happening *during* the flight of the particles from the source to the screen;
(b) assumes the existence of the trajectories; and hence,
(c) worries about the causality underlying the formation of those trajectories.

The prominent example of such a perspective is that provided by the de Broglie–Bohm pilot wave interpretation, which assumes an *ontic* role of the wave function.

As mentioned in Subsection 8.1.3, the pilot wave interpretation considers the wave function ψ as representing a dynamical guiding field for the motion of the individual particles, thereby supporting the notion of a 'trajectory'. Thus, from an ontological perspective, the function $|\psi|^2$, in the case of electrons, could be interpreted as the probability density for an electron *being* (rather than *found* if looked for, [2]) in an infinitesimal length element around a future position (on the screen). And, in an analogous manner, in the case of agents, such a function would then give the probability density for an agent *being* in an infinitesimal element around a potential future price.

Following [3], in the present chapter we make use of both the Einstein–Born–Ballentine 'ensemble' interpretation and the de Broglie–Bohm 'pilot-wave' interpretation of the wave function.

By following Sections 9.1 and 9.3, in the next two sections, we provide a quantum dynamical analysis of the electron two-slit interference and hence, of the agent two-preference interference in an analogous manner. In the usual analysis of the former, electrons flying from the source to the screen via the slits are treated as free particles (i.e. $V = 0$). However, we not only discuss this case (see Section 10.1), but also consider the situation in which the motion of the electrons toward the screen is influenced by some potential (Section 10.2). As the simplest case, we take the corresponding potential energy to be linearly depending on the particle (electron/agent) position (i.e., $V = -\mathbf{K} \cdot \mathbf{x} = -Kx$ in one dimension). In the case of the agents, the position-like variable (i.e. 'x') represents agent preference and hence, in this way, the finance analogue of the case of nonzero potential enables a physics-like comprehension of the influence of the social (economic/financial) environment on the preferences of the agents.

Continuing from Section 10.3 that involves a discussion on the assumptions that are commonly applicable to the dynamical models of both electrons and agents (owing to the precise mapping between them), Chapter 11 provides simulations (in terms of different values of the parameter K) of the probability density functions as well as of the particle trajectories. It also explores the physics-like interpretations of the agent two-preference interference model parameters.

10.1 Electrons versus Agents: $V = 0$

10.1.1 Electron Two-Slit Interference

As already discussed, the experimental setup in this case [Figure 9.1] involves a source, a pair of slits and a screen. The motion of the electrons – from the source to the screen via the slits can be described with reference to a two-dimensional Cartesian system: with the horizontal x-axis and the vertical y-axis. The wave function for an electron travelling from the source to the slits is that of a plane wave, given by

$$\psi(x, y) = a\, e^{i(k_1 x + k_2 y)}, \tag{10.1}$$

where a is a constant. The y-propagation of the plane wave remains unaffected throughout the ('source to screen via slits') journey of the electron. Therefore, we consider only the x-component of the electron motion. Also, for the plots of the probability density functions and the trajectories shown in this chapter, the vertical axis is used to label the time parameter.

We assume that at $t = t_{initial} = 0$ the slits produce waves having identical Gaussian profiles along the x-axis and centred at $\pm X$. We also assume that the corresponding wave packets have equal initial width ρ and equal and opposite group velocities having x-components $+u$ and $-u$). Following [4], the initial forms of the two waves coming out from the slits are thus given by

$$\psi_-(x, 0) = \rho^{-1/2} \exp\left[-\frac{(x+X)^2}{2\rho^2} + \frac{im}{\hbar}(+ux)\right], \text{ and} \tag{10.2}$$

$$\psi_+(x, 0) = \rho^{-1/2} \exp\left[-\frac{(x-X)^2}{2\rho^2} + \frac{im}{\hbar}(-ux)\right]. \tag{10.3}$$

Ignoring the 'y-propagation' plane wave component, the wave function ψ_- is effectively a packet moving along the x-axis in the $+x$-direction with group velocity $+u$. The suffix '$-$' in ψ_- indicates that the wave function corresponds to the slit on the negative side of the x-axis. Similarly, the wave function ψ_+ is effectively a packet corresponding to the slit on the positive side of the x-axis and moving along the $-x$-direction with group velocity $-u$.

At time t the total (nonnormalized) wave function – as the solution to Equation (2.35), with $V = 0$ – is given by

$$\psi(x, t) = \psi_-(x, t) + \psi_+(x, t) \tag{10.4}$$

where

$$\psi_-(x,t) = \rho_t^{-1/2} \exp\left[-\frac{(x+X-ut)^2}{2\rho\rho_t} + \frac{im}{\hbar}(+u)\left(x - \frac{ut}{2}\right)\right], \text{ and} \tag{10.5}$$

$$\psi_+(x,t) = \rho_t^{-1/2} \exp\left[-\frac{(x-X+ut)^2}{2\rho\rho_t} + \frac{im}{\hbar}(-u)\left(x + \frac{ut}{2}\right)\right] \tag{10.6}$$

with

$$\rho_t = \rho\left(1 + \frac{i\hbar t}{m\rho^2}\right). \tag{10.7}$$

By expressing the total wave function (10.4) in polar form, we can equate it with $\psi(x,t) = R(x,t)\exp iS(x,t)/\hbar$ to identify the amplitude $[R(x,t)]$ and the phase $[S(x,t)]$ functions. Using $R(x,t)$ we can obtain $\psi^*\psi = |\psi|^2 = R^2(x,t)$ as the probability density function. Using the phase function $S(x,t)$, the particle trajectories can be obtained by integrating

$$m\dot{x} = \frac{\partial S}{\partial x} \tag{10.8}$$

for different initial positions, $x_0 = x(t = t_{initial} = 0)$, lying within the slits. Although the classical potential energy is zero, the spatial variation of the gradient of the amplitude of the total wave function gives rise to the quantum potential energy, given by

$$Q = -\frac{\hbar^2}{2m}\frac{\partial^2 R}{\partial x^2}. \tag{10.9}$$

As discussed in Section 2.8, with respect to the de Broglie–Bohm pilot-wave perspective, it can be said that, due to Q there is a 'quantum' force acting on the electrons that takes them from their initial positions at the slits to the final positions on the screen.

All these expressions (i.e., of R, S and Q) have complicated forms [4]. We avoid stating those here, as their explicit expressions provide no additional insights that would be useful for further discussions and analyses. In the following we proceed to develop a dynamical model of the case of agents in analogy with that for the electrons.

10.1.2 Agent Two-Preference Interference

We begin by revisiting the financial analogue of the idea of the *source* as discussed in Section 9.3.1. For the case of electrons, we recall that the source is to be considered as an ensemble of identically prepared identical systems. Analogously, with reference to our discussion from the previous chapter, for an economic/financial process, the source can be understood as an ensemble of individual agent-systems participating in the same context (transaction trading).

To enable a position-like representation of the wave function of an agent, we introduce a $(2+1)$-dimensional *preference* space-time $\mathbb{P}$, having two spatial and one temporal dimensions. The structure of $\mathbb{P}$ is assumed to be Cartesian and the points belonging to such a space are the ordered pairs, that is, $(x, y, t) \in \mathbb{P}$. The motion of an agent travelling from the source to the slits will be described by a plane wave, $\psi = a\, e^{i(k_1 x + k_2 y)}$, where a is a constant.

The pair of base-preferences in the 'agents' case plays the role of the pair of slits in the 'electrons' case, with the centres of the slits located at the points $(\pm X, 0)$. As already discussed, the base preferences are to be considered as mutually orthogonal, in the sense that these preferences correspond to two mutually exclusive intentions – they want to trade at a potential future price that is either *higher* or *lower* than the previous one – for the individual agents. In analogy with the case of 'electrons', we can consider finite but equal numerical ranges for each of these base preferences. These ranges will then work as the 'preference confinements', to both the sides of $x = 0$, for the individual agents participating in the trade, in analogy with the 'position confinements' for the individual electrons and will thus enable an electron-like wave packet representation of the agent-states. The screen in this case is just a metaphor for the time instant at which we wish to obtain the distribution of the probability density over the agent preferences.

Similar to the 'electrons' case, we ignore the y-propagation of the wave and therefore, the 'agent preference', playing the role analogous to the 'electron position', can be defined by the x-variable alone and be given by the 1D agent-position $x(t) \in \mathbb{P}$ (see Section 9.2.1) as

$$x(t) = \frac{p(t) - p(t_k)}{p(t_k)}. \tag{10.10}$$

The preceding equation requires some clarification about the time-labels (see Figure 10.1):

- For $t = t_k$, $x(t_k) = 0$. Here, $t_k < 0$ is the instant at which the previous stock share price has already been realized. This gives the source location, that is, $x = 0$.
- For $t = 0 = t_{initial}$,

$$x(0) = \frac{p(0) - p(t_k)}{p(t_k)}. \tag{10.11}$$

This instant is an intermediate one between the instants t_k and t_{k+1}, that is, $t_k < t_{initial} < t_{k+1}$. This instant marks the arrival of the agents at the slit positions and hence corresponds to the formation of their initial preferences.

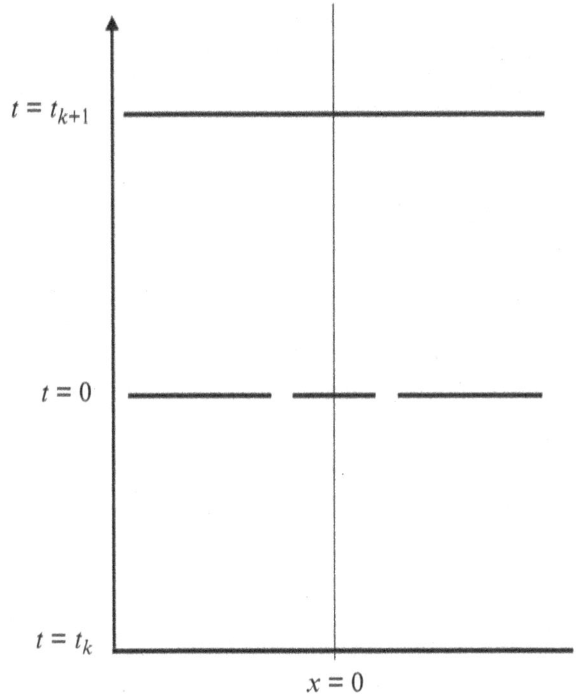

Figure 10.1 About the time labels: $t = t_k \longrightarrow t = 0 \longrightarrow t = t_{k+1}$.

- For $t = t_{k+1}$, where, $t_{k+1} > 0$ (indicating the future time instant),

$$x(t_{k+1}) = \frac{p(t_{k+1}) - p(t_k)}{p(t_k)}. \tag{10.12}$$

  The t_{k+1} instant is the time instant from the future for which the model is expected to predict a distribution of probability density over a range of potential stock share prices. This instant marks the arrival of the agents at different screen positions.[3]

Thus, from Equation (10.10), for example, it follows that $x(t = t_{k+1}) < 0$ implies $p(t_{k+1}) < p(t_k)$, that is, 'the potential future price less than the previous one'. And, similar to the 'electrons' case, this time we can take (the same) Equation (10.4) as the time-evolved *agent*-state given in terms of the superposition of two wave functions, $\psi_-(x, t)$ and $\psi_+(x, t)$, which now (i.e. in the 'agents' case) correspond to the agents' intention to transact, respectively, at a price *lower* than and *higher* than the previous price.

[3] With reference to the from-via-to process discussed in Section 8.2, in the 'agents' case, the source, slits and screen correspond to the instants $t_{from} = t_k$, $t_{via} = t_{initial}$ and $t_{to} = t_{k+1}$, respectively.

10.2 Electrons versus Agents: $V = -Kx$

In the previous section we considered the *free* (i.e. with $V = 0$) 'slit-to-screen' dynamics for the electron two-slit interference. Also, in an analogous manner we assumed the *free* evolution of the preferences of the agents (i.e. without any influence of their social surroundings). It is well known that such an analysis, for the 'electrons' case, does lead to an explanation of the experimentally observed electron intensity distribution. However, it does not let us explore the behaviour of the agents (again, in terms of the evolution of their preferences) under the influence of the surrounding complex social / economic / financial environments.

Therefore, for a more realistic analysis, the agent two-preference interference model – involving the behaviour of agents in a financial trading scenario – should necessarily take into account the influence of the financial environment on the preferences of the individual agents. Indeed, there is a huge literature talking about how the socio-economic environment of the agents could influence their choices and preferences.

As discussed previously, in a physics-like model such an influence would be expressed in terms of a function analogous to that of a potential energy function. Now, there is a broad variety of mathematical forms for the potential energies as functions of the particle positions in physics and it is indeed hard to determine which of those forms could be appropriate for analogously capturing the economic/financial influences on the agent behaviour.

Here, we choose the simplest form for the classical potential energy: $V(x) = -Kx$, that is, it changes linearly with the particle position.[4]

In this case, the initial forms of the two waves coming out from the slits are the same as those for the $V = 0$ case, that is, they are given by Equations (10.2) and (10.3).

However, for $t > 0$ the total (nonnormalized) wave function – as the solution to Equation (2.35), with $V = -Kx$ – is given by [4] as follows:

$$\psi(x,t) = \psi_-(x,t) + \psi_+(x,t), \tag{10.13}$$

where

$$\psi_-(x,t) = \rho_t^{-1/2} \times \exp\left[-\frac{(x + X - ut + \frac{Kt^2}{2m})^2}{2\rho\rho_t} + \frac{im}{\hbar}\left(+u - \frac{Kt}{m}\right)\left(x - \frac{ut}{2}\right) - \frac{iK^2t^3}{6\hbar}\right], \text{ and} \tag{10.14}$$

[4] The most familiar example of this type of dependence is the gravitational potential energy: $V(x) = -mgx$. The reader may recall the discussion from Section 2.8.

$$\psi_+(x,t) = \rho_t^{-1/2} \times \exp\left[-\frac{(x - X + ut + \frac{Kt^2}{2m})^2}{2\rho\rho_t} + \frac{im}{\hbar}\left(-u - \frac{Kt}{m}\right)\left(x + \frac{ut}{2}\right) - \frac{iK^2t^3}{6\hbar}\right], \tag{10.15}$$

with $\rho_t = \rho\left(1 + \frac{i\hbar t}{m\rho^2}\right)$.

It can be easily seen that Equations (10.14) and (10.15) reduce to Equations (10.5) and (10.6) for $K = 0$.

Similar to what was discussed at the end of the previous section, using the polar form of the wave function from Equation (10.13), the amplitude function $[R(x,t)]$, the phase function $[S(x,t)]$ and hence, the probability density (R^2), the quantum potential energy Q and the trajectory equation (for $x(t)$) can be obtained [4].

In the next chapter, we turn to the analysis of a variety of plots for the distribution of probability density, the quantum potential energy, and the particle trajectories.

10.3 A Word on the Model Assumptions

An essential requirement of the approach to the systematic construction of a model of reality in theoretical physics is the clarity of the 'assumptions' of the model. In physics, often the main purpose of making assumptions is to make the model of reality simple to begin with and hence, easily solvable. For example, when 'projectile motion' is introduced to high school students, often the construction of the corresponding dynamical model is shown by first *neglecting* the air resistance to the motion of the projectile, so that, in the dynamical analysis of the motion, we only have to worry about the force due to gravity alone.

Although *in reality* the air resistance is always present, when it comes to modelling the dynamics of the projectile motion, it makes more sense to ignore its effect and get an idea of the (simplified) physics of the situation, than to start directly by involving effects such as the Stokes drag or the Newton drag and getting into the realistic but more complicated physics of the situation (see, e.g., [5] for details). Our understanding of the physical behaviour of reality through its model is thus subject to the assumptions of the model.

As is often the case, in physics, if you relax a few simplifying assumptions of the previous (less accurate) model, the resulting model provides you with a more accurate representation of the reality by involving more general mathematical relationships between physical quantities. For instance, the 'special relativistic' mechanical model of reality is certainly a more accurate representation of reality than that offered by the 'Galilean–Newtonian' mechanical model. Having said this, it is also important to understand that the 'Galilean–Newtonian' mechanical model cannot be regarded as an 'untrue' representation of reality. It works very well in the 'domain of applicability' where relativistic effects (i.e. those arising due to extremely high speeds – close to the speed of light – of the objects) are ignorable.

Thus, even in the case of a physics-like model of an economic or financial reality, we will need to keep in mind that the understanding of the reality as offered by such a model is subject to its assumptions, and hence, it may not provide a very accurate representation of the reality that the model refers to. Nevertheless, for a systematic use of a physics-analogous way of modelling economic/financial reality, we think it is necessary to begin with a simpler (or even, an oversimplified) model. As the example of 'projectile motion' implies, the clarity of the assumptions of such a relatively simpler model will then serve as a clue for extending the model and replacing it by a more complicated (hence, perhaps, a more realistic) one.

So let us take a review of the assumptions of the dynamical models given in the previous two sections. The assumptions are as follows[5].

1. As remarked at the beginning of the present chapter, in analogy with the individual electrons, we consider individual agents as single-particle systems. (In all of the following plots the electrons and the agents are commonly labelled as particles.)
2. The slit widths in the case of 'electron two-slit interference' to both the sides of $x = 0$ are assumed to be equal. Analogously, the preference ranges in the case of 'agent two-slit interference' to both the sides of $x = 0$ are also assumed to be equal. The waves coming out of both slit widths (in the 'electrons' case) and preference ranges (in the 'agents' case) are assumed to have a Gaussian profile. (In the captions for the following plots, 'ρ' denotes both the slit widths and the preference ranges.)
3. The slit centres on the x-axis in the 'electrons' case are assumed to be symmetrically located about $x = 0$. Analogously, the centres of the preference ranges on the x-axis in the 'agents' case are also assumed to be symmetrically located about $x = 0$. (In the captions for the following plots, 'X' denotes the numerical value of both the slit centres and the centres of the preference ranges.)
4. In the cases of both electrons and agents, the group velocities of the wave packets are assumed to be equal and opposite. (In the captions for the following plots, 'u' denotes the numerical value of the group velocities in both cases.)
5. The influence of the economic/financial environment on the agent preferences is assumed to be given by a potential energy-like function, similar to the way in which the influence on the particle positions is modelled in terms of the potential energy functions from physics.

[5] Here we simply state the qualitative assumptions and the assumptions pertaining to the model parameters. The interpretations of the latter are explored in Section 11.2.

6. As the simplest mathematical form, the potential energy function is taken as linearly depending ($V = -Kx$) on the positions of the electrons. The same function is assumed as representing the influence of the economic/financial environment on the agent preferences in the trading scenario, with K being expected to have a force-like interpretation.

References

[1] Peres, A. (2002). *Quantum Theory: Concepts and Methods*. Dordrecht: Kluwer Academic Publishers.
[2] Norsen, T. (2017). *Foundations of Quantum Mechanics: An Exploration of the Physical Meaning of Quantum Theory*. Cham: Springer.
[3] Athalye, V. (2025). Unifying Perspectives: All-encompassing Reality vs. All-inclusive Physics. *Journal of Physics: Conference Series*. 2948: n. pag.
[4] von Bloh, K. and Athalye V. 2022. The Double-Slit Experiment in a Uniform Potential. *Wolfram Demonstrations Project*. https://demonstrations.wolfram.com/TheDoubleSlitExperimentInAUniformPotential/.
[5] Greiner, W. (2004). *Classical Mechanics: Point Particles and Relativity*. New York, NY: Springer.

11

Simulations and Interpretations

With the background of the assumptions from the previous chapter, we now turn to the simulations and the interpretations of the model parameters. The simulations given in this chapter follow from the calculations of the trajectories $x(t)$, the probability density R^2 and the quantum potential Q (abbreviated as QP in the plots), which are based on the code by Klaus von Bloh and Vikram Athalye [1].[1] As already discussed, we will interpret the (electron / agent) wave function in two manners: the *epistemic* one following from the Einstein–Born–Ballentine 'statistical' interpretation and the *ontic* one following from the de *Broglie–Bohm* 'pilot wave interpretation. Therefore, in this section, we will be concerned specifically with the plots of the probability density R^2 and of the trajectories $x(t)$, in different conditions.

11.1 Simulations of Probability Distributions and Trajectories

The function $x(t)$, plotted for a range of initial conditions, gives different possible single-particle trajectories. The analogue of the ('within slits') initial positions for the case of the 'electrons' is the (within base preferences) initial quotes for the case of the 'agents'. And, the financial analogue of the idea of the 'electron trajectories' starting from those initial positions is that of the 'agent (or the market participant) trajectories' starting from the initial quotes. For the electrons the particle trajectories belong to the position configuration space, whereas for the agents these analogously belong to the preference configuration space. The plot of the function $R^2(x, t)$ gives the distribution of probability densities for the x variable. The probability density, that is, the probability per unit length (in one dimension), is proportional to the number of particle trajectories per unit length.

[1] It is available at https://demonstrations.wolfram.com/TheDoubleSlitExperimentInAUniformPotential/. The plots of R^2 and Q (i.e. QP) are scaled to fit. The wave function used is not normalized. Also, for all the plots: $m = \hbar = 1$.

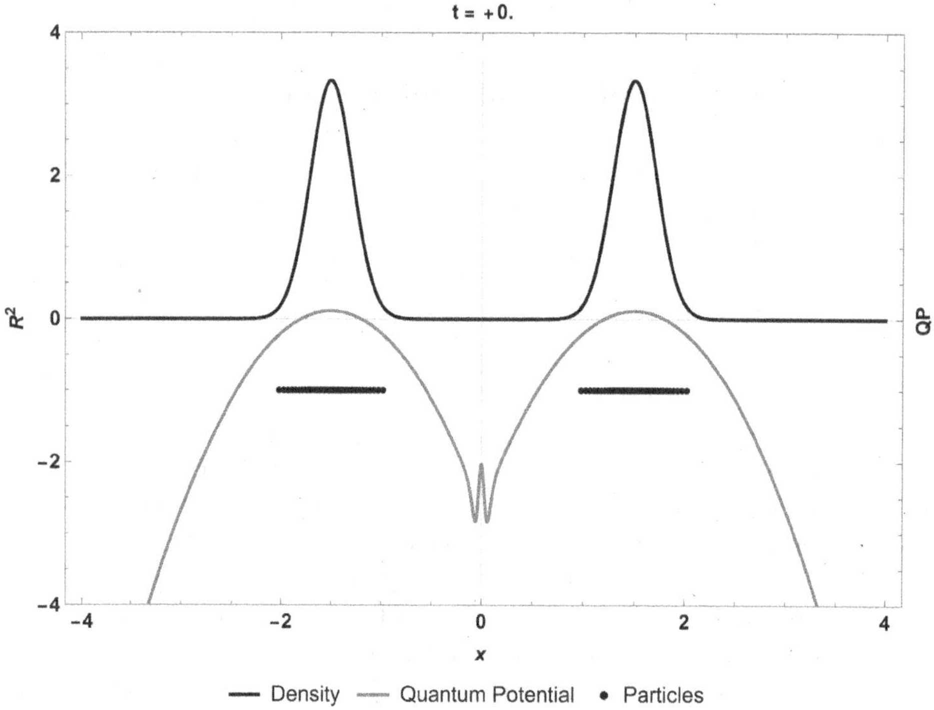

Figure 11.1 Cross-sectional plots of the initial probability density (R^2) and quantum potential (QP) at the slits ($t = 0$), along with particle positions for $X = 1.5$, $\rho = 0.3$ ([1]). The 'thick' horizontal black segments, symmetrically located at both the sides of $x = 0$, indicate slits, with centres at $X = \pm 1.5$.

Figure 11.1 gives the cross-sectional plot of the initial probability density that is, $R^2(t = 0)$ and also, of the quantum potential (QP) at the slits at $t = 0$. This way, the R^2 plot indicates the modelling of the experimental 'initial state preparation' ('Gaussian' profiles for the wave functions at the slits). It is independent of the value of the parameter K (from the expression $V = -Kx$) and hence, is the same for all the remaining plots.

These remaining plots can be grouped with respect to the parameter K as follows:

- Case – 1: $K = 0$: Figures 11.2 and 11.3.
- Case – 2: $K \neq 0$: Figures 11.4, 11.5; 11.6, 11.7; 11.8, 11.9; 11.10, 11.11.
- Case – 3: $K \ll 0$ and $K \gg 0$: Figures 11.12, 11.13; 11.14, 11.15; 11.16, 11.17.

We now proceed to carry out the *observations* on the plots from the above cases. The *interpretations* that may follow these observations are discussed in Section 11.2.

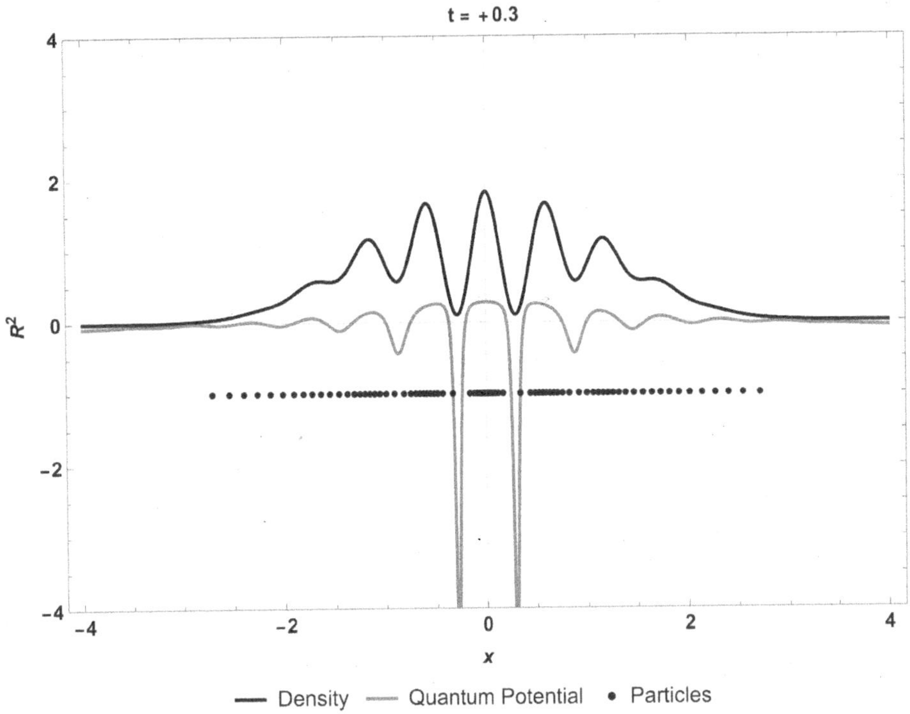

Figure 11.2 Cross-sectional plots of the initial probability density (R^2) and quantum potential (QP) ($t = 0.3$), along with particle positions for $X = 1.5,\ \rho = 0.3$ ([1]).

11.1.1 Case – 1: $K = 0$

Figure 11.2 shows the plots of the probability density and the quantum potential for the time parameter value $t = 0.3$, whereas Figure 11.3 shows the particle trajectories $x(t)$ formed over the time parameter range $0 < t < 0.6$, for the values of the parameters X, u, ρ and K as given in the caption. These plots correspond to the *usual* electron two-slit interference, in the sense that the electrons freely move from the slits to the screen, that is, with $V = 0$ in the Schrödinger equation. It is observed that the plots of the probability density function and the quantum potential, and also, the distribution of the trajectories are all symmetric about $x = 0$. The probability density plot matches with the intensity pattern, which would be built by sending electrons one-by-one from the source – through the slits – to the screen of a double-slit apparatus ([2], [3]).

In an analogous manner, these plots represent the agent two-preference interference situation as well. Thus, the trajectories of Figure 11.3 can also be seen as displaying the evolution of the agent preferences from their initial values (at $t = 0$) to the final values (at $t = 0.6$). The plot of R^2 shown in Figure 11.2, when

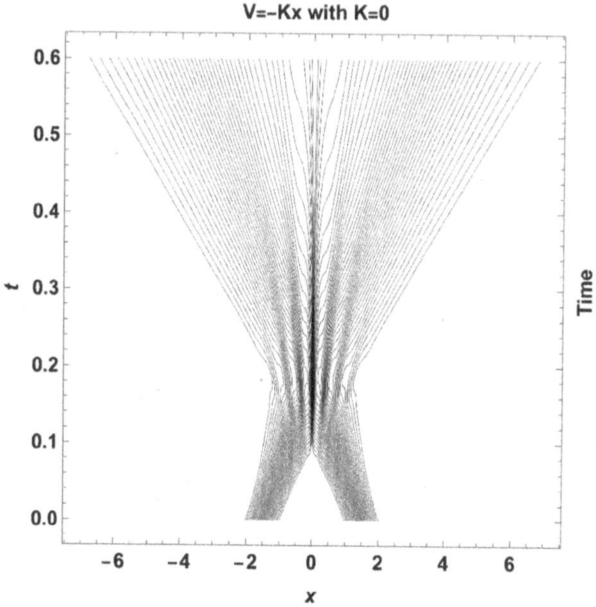

Figure 11.3 Trajectories for two Gaussian wave profiles, with a Gaussian distribution of initial positions at each slit for $V = -Kx$ and $K = 0,\ X = 1.5, u = 8,$ $\rho = 0.3$ ([1]).

considered with respect to the case of 'agents', can be said to give the distribution of probability density over the agent preferences for the future stock share prices. The peak of the R^2 distribution is observed at '$x = 0$', for which the density of the trajectories from Figure 11.3 can also be seen as the highest.

11.1.2 Case – 2: $K \neq 0$

The observations stated in this and the next subsection refer to the case of 'agents'. In other words, we avoid making a constant reference to the 'electrons' case in these subsections, as we hope that by now, it is clear to the reader that the dynamical model for the 'agents' case is developed by drawing an analogy from the dynamical model for the 'electrons' case.

Figures 11.4, 11.6, 11.8 and 11.10 show the plots of probability densities as the functions of the agent-preferences (and hence, of the 'potential' future stock share prices) and also, of the quantum potential energy-like functions for $V = -Kx$, with different (nonzero) values of K, whereas Figures 11.5, 11.7, 11.9 and 11.11 show the distribution of the agent-trajectories for those same values of K, with the other parameter values given in the respective captions.

For $K = +5$ Figure 11.4 shows a shift (w.r.t. $x = 0$) in the distribution of the probability density in the $x < 0$ region, whereas for $K = -5$ Figure 11.8 shows a

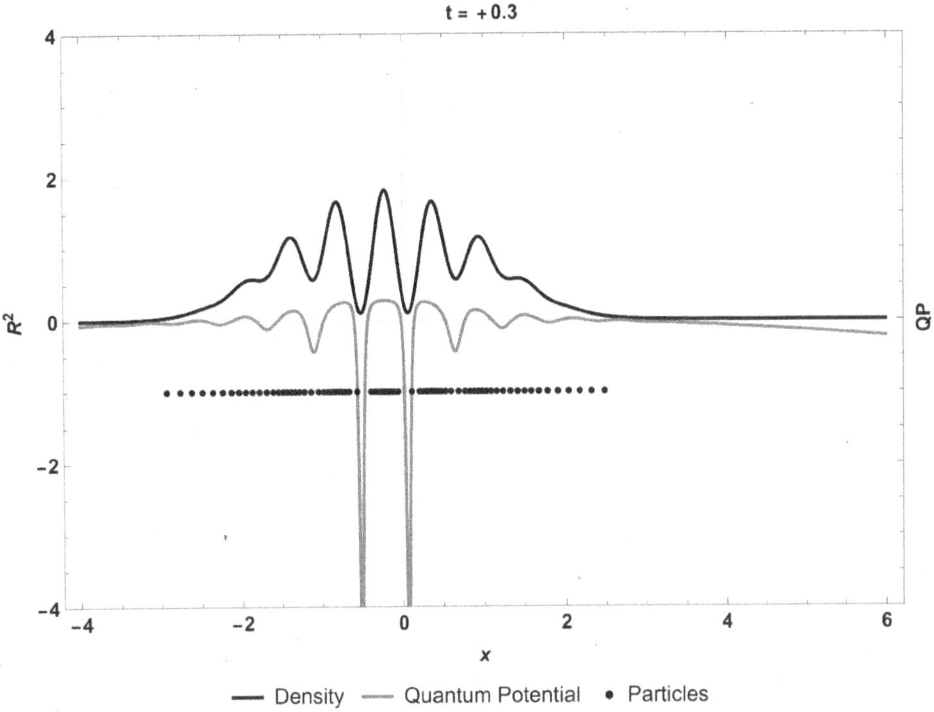

Figure 11.4 Cross-sectional plots of R^2 and Q along with particle positions ($t = 0.3$) for $V = -Kx$ and $K = +5$, $X = 1.5$, $u = 8$, $\rho = 0.3$ ([1]).

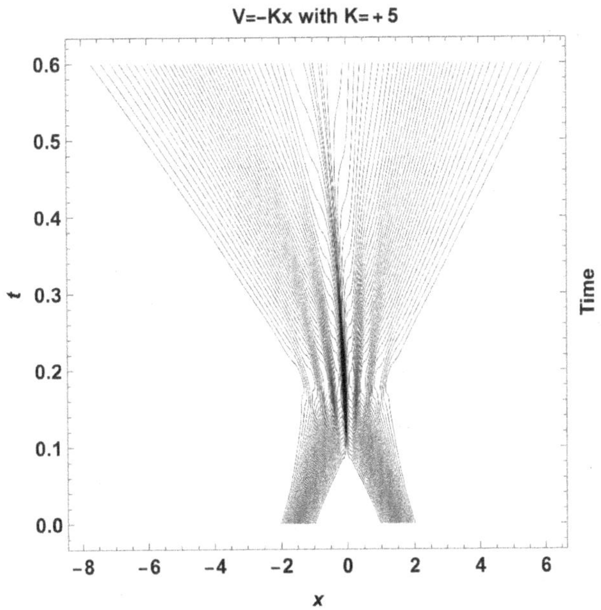

Figure 11.5 Trajectories for $V = -Kx$ and $K = +5, X = 1.5, u = 8, \rho = 0.3$ ([1]).

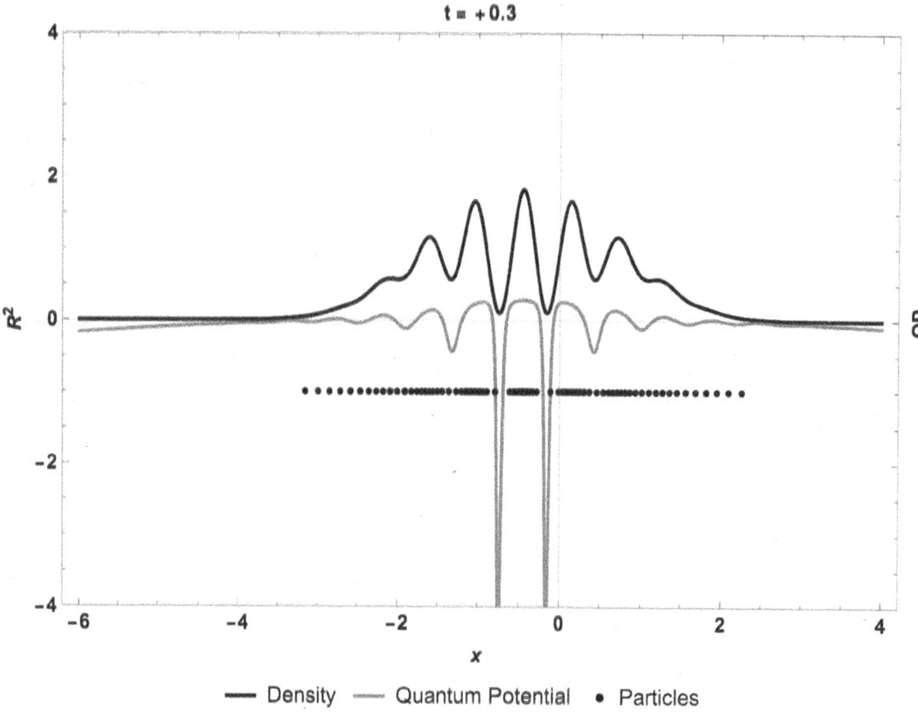

Figure 11.6 Cross-sectional plots of R^2 and Q along with particle positions ($t = 0.3$) for $V = -Kx$ and $K = +10$, $X = 1.5$, $u = 8$, $\rho = 0.3$ ([1]).

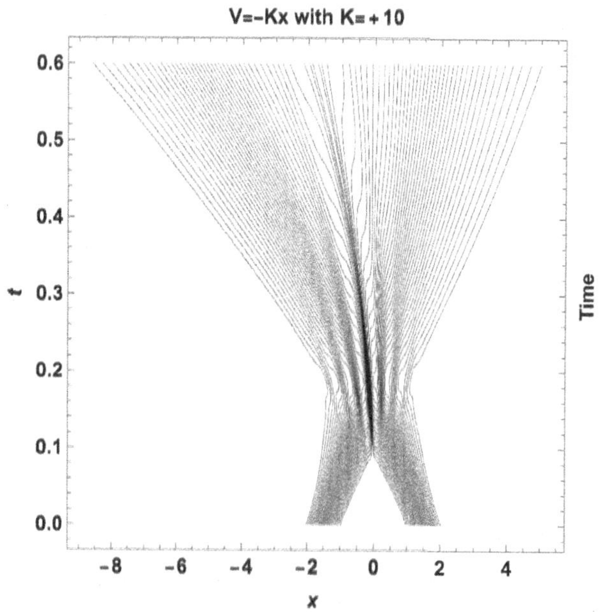

Figure 11.7 Trajectories for $V = -Kx$ and $K = +10$, $X = 1.5$, $u = 8$, $\rho = 0.3$ ([1]).

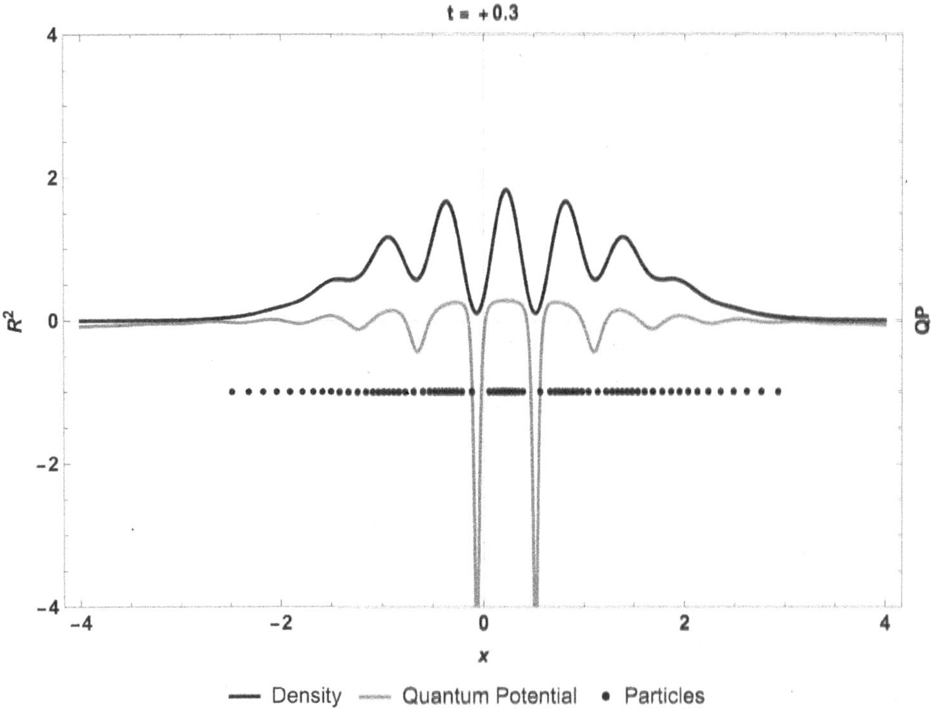

Figure 11.8 Cross-sectional plots of R^2 and Q along with particle positions ($t = 0.3$) for $V = -Kx$ and $K = -5$, $X = 1.5$, $u = 8$, $\rho = 0.3$ ([1]).

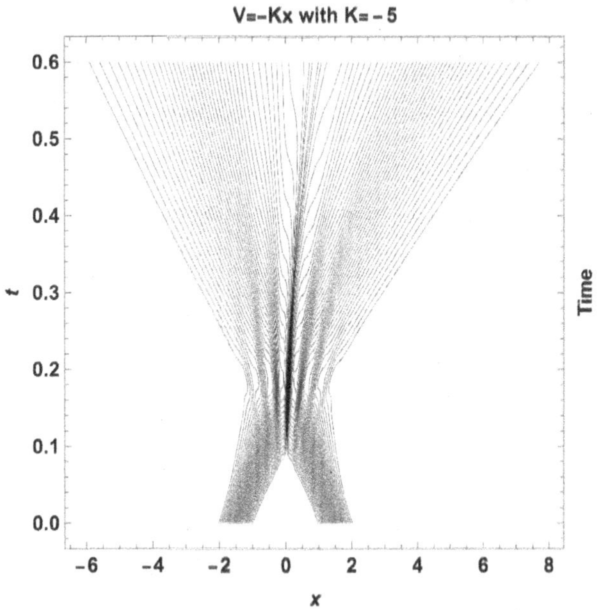

Figure 11.9 Trajectories for $V = -Kx$ and $K = -5, X = 1.5, u = 8, \rho = 0.3$ ([1]).

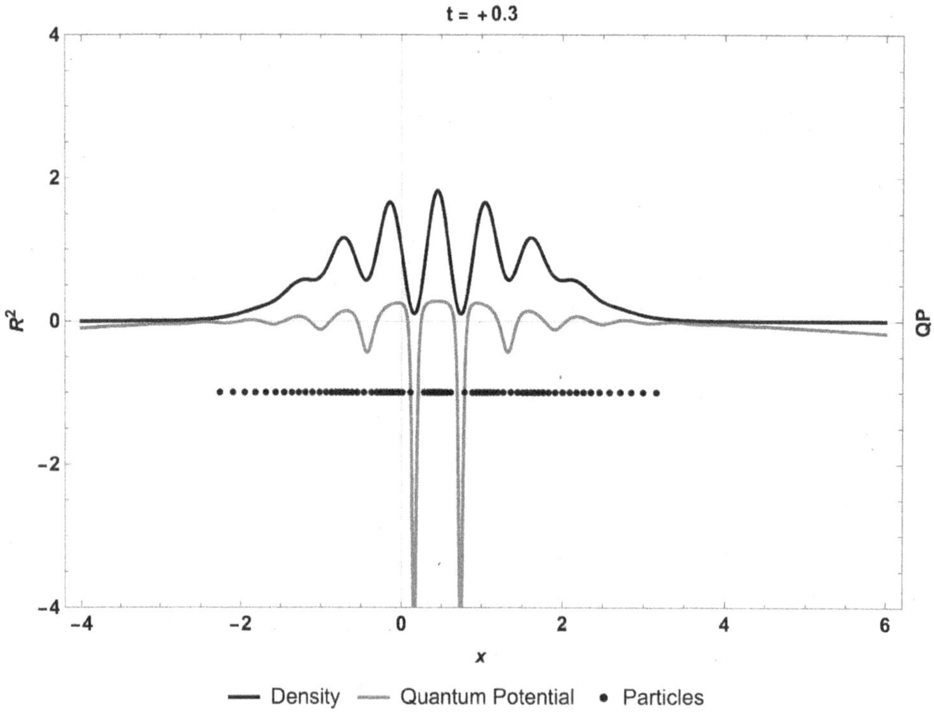

Figure 11.10 Cross-sectional plots of R^2 and Q along with particle positions ($t = 0.3$) for $V = -Kx$ and $K = -10$, $X = 1.5$, $u = 8$, $\rho = 0.3$ ([1]).

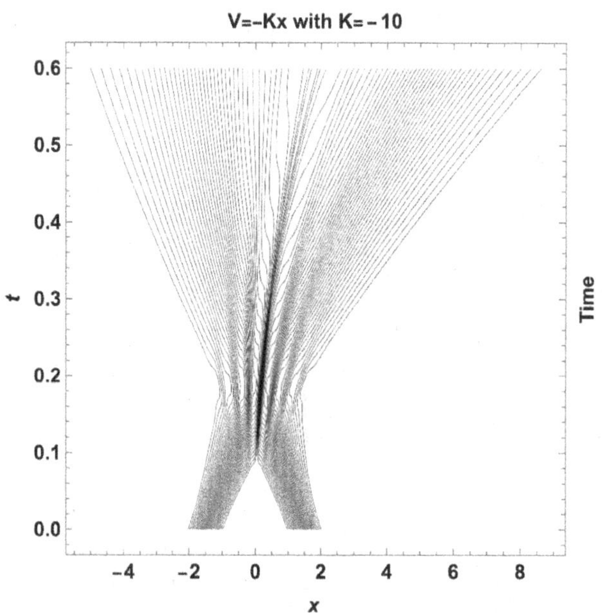

Figure 11.11 Trajectories for $V = -Kx$ and $K = -10$, $X = 1.5$, $u = 8$, $\rho = 0.3$ ([1]).

shift (w.r.t. $x = 0$) in the distribution of the probability density in the $x > 0$ region. Figures 11.6 and 11.10 show further shifts, respectively, in the $x < 0$ and the $x < 0$ regions due to an increase in the strength of the K-parameter ($K = \pm 10$).

Owing to the influence of $V = -Kx$ (and therefore, of K), in Figures 11.5, 11.7, 11.9 and 11.11, it can be observed that these *shifts* in the distributions of the probability densities over the potential future stock prices directly correspond to the *bending* of the agent-trajectories (in the $x < 0$ and the $x > 0$ regions) in the respective cases. Also, in all these figures, the *peaks* of the probability distributions for the potential future stock prices and the *highest densities* of the agent-trajectories can be consistently observed around the same locations on the x-axis, that is, around the same values of the agent-preferences.

11.1.3 Case – 3: $K \ll 0$ *and* $K \gg 0$

The cases discussed in the present subsection refer to extreme conditions (see Figures 11.12, 11.13; 11.14, 11.15), that is, for very large (numerical) values of the K-parameter. In Figures 11.12 and 11.14 we can observe *extreme shifts* in the distributions of the probability densities over the potential future stock

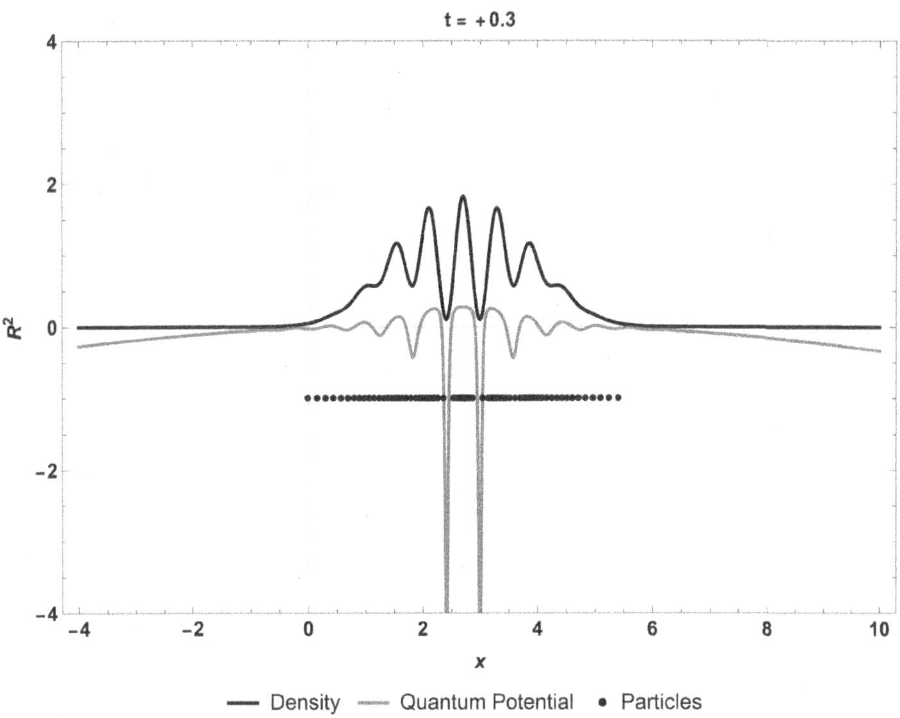

Figure 11.12 Cross-sectional plots of R^2 and Q along with particle positions ($t = 0.3$) for $V = -Kx$ and $K = -60$, $X = 1.5$, $u = 8$, $\rho = 0.3$ ([1]).

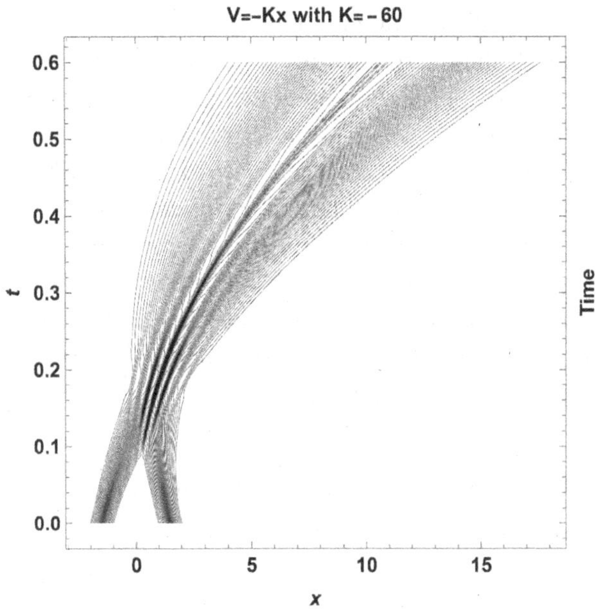

Figure 11.13 Trajectories for $V = -Kx$ and $K = -60$, $X = 1.5$, $u = 8$, $\rho = 0.3$ ([1]).

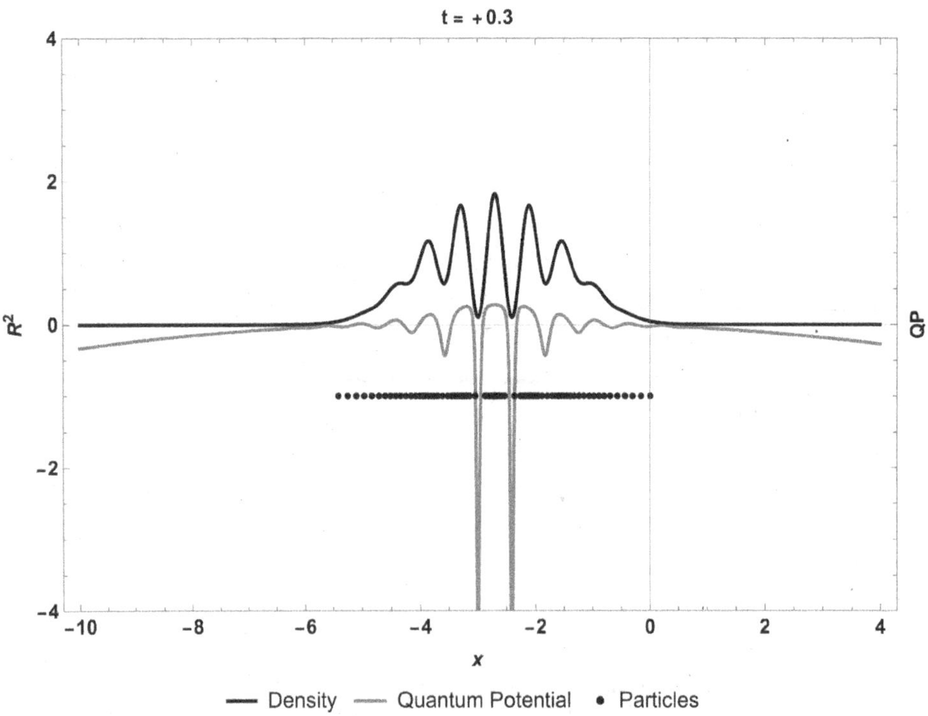

Figure 11.14 Cross-sectional plots of R^2 and Q along with particle positions ($t = 0.3$) for $V = -Kx$ and $K = +60$, $X = 1.5$, $u = 8$, $\rho = 0.3$ ([1]).

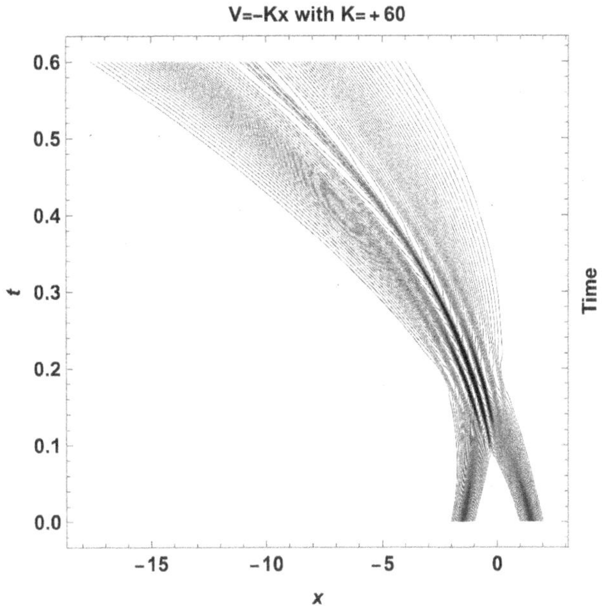

Figure 11.15 Trajectories for $V = -Kx$ and $K = +60$, $X = 1.5$, $u = 8$, $\rho = 0.3$ ([1]).

prices, respectively, in the $x > 0$ and the $x < 0$ regions. Also, the peaks of the probability density functions can be seen to have moved at the locations that are much beyond the initial preference-ranges (slits) in those regions. Correspondingly, we can also observe an *extreme bending* of the agent-trajectories (in the $x < 0$ and the $x > 0$ regions) in Figures 11.13 and 11.15, with, again, the *highest densities* of the agent-trajectories having locations that are much beyond the initial preference-ranges.

Moreover, it is interesting to observe that the agent-trajectories, originating from the 'positive-side' and the 'negative-side' preference-ranges appear to 'get together and start bending' relatively *earlier* as shown in Figures 11.16 and 11.17 than that shown in Figures 11.13 and 11.15) when the value of the parameter X is *reduced* (from 1.5 to 1.0).

With the background of all these simulations, in the next section, we propose the interpretations of the parameters used in the agent two-preference interference model.

11.2 Interpretations of the Model Parameters

As seen in the preceding section, the dynamical model for the agent two-preference interference (as an imitation of the dynamical model for the electron two-preference interference) involves the following parameters:

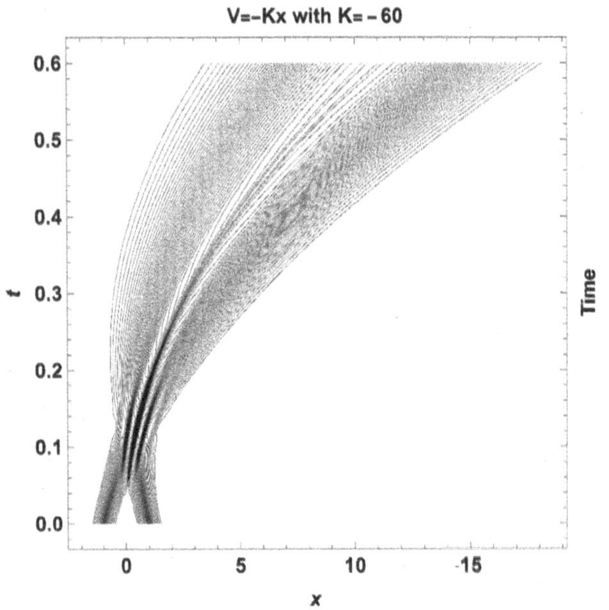

Figure 11.16 Trajectories for $V = -Kx$ and $K = -60$, $X = 1.0$, $u = 8$, $\rho = 0.3$ ([1]).

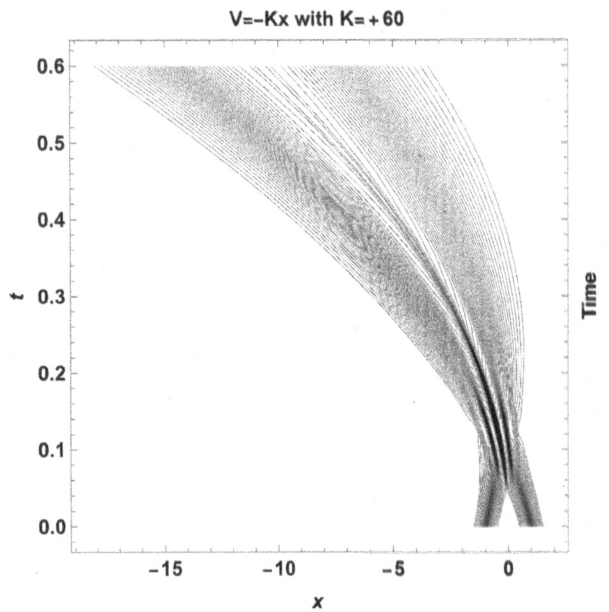

Figure 11.17 Trajectories for $V = -Kx$ and $K = +60$, $X = 1.0$, $u = 8$, $\rho = 0.3$ ([1]).

- initial preference-ranges (ρ),
- numerical value of the means of the initial preference-ranges (X), and
- the K-parameter and the quantum potential energy-like function (Q).

In the following, apart from the possible (physics-analogous and finance-relevant) interpretations of these parameters, we also comment on the expectation values of the potential future stock share price and the corresponding agent-preference.

11.2.1 Interpretations of X and ρ

Let us consider each of the (initial) agent wave functions, $\psi_-(x)$ and $\psi_+(x)$, as representing a subcollection of both, informed and uninformed traders. One of the assumptions discussed in Section 10.3 refers to the form of the waves originating from the preference-ranges: these have been assumed to have a Gaussian profile. Due to such an assumption, the expectation values of the initial agent-preferences are the same as the arithmetic 'means' of the preference-ranges (in analogy with the slit-centres in the 'electrons' case). In a typical trading scenario, the initial quotes could be due to the overreacting traders or the underreacting traders[2]. We can therefore interpret the coordinates of the means (i.e. $\pm X$) of the initial preference-ranges, in terms of the following implications:

1. Overreacting traders $\Longrightarrow$ initial quotes with relatively *larger* difference between the previous and the (potential) next stock share prices $\Longrightarrow$ $\pm X$ away from $x = 0$.
2. Underreacting traders $\Longrightarrow$ initial quotes with relatively *smaller* difference between the previous and the (potential) next stock share prices $\Longrightarrow$ $\pm X$ closer to $x = 0$.

Also, the initial preference-ranges (ρ), in analogy with the slit-widths, can be interpreted as the 'uncertainties' in the anticipations of the agents (traders) pertaining to the potential future stock share price. Thus, it may be argued that a relatively smaller value of ρ would amount to assuming that the agent subcollections are dominated by informed traders, whereas a relatively larger value of ρ would amount to assuming that the agent subcollections are of predominantly uninformed traders.

11.2.2 A Comment on the Expectation Values: $\langle x(t)\rangle$ and $\langle p(t)\rangle$

The appearance of the peak of the distribution in Figure 11.2 (for the $K = 0$ case) leads to a novel insight: the agent two-preference model (subject to the assumptions

[2] Such reactions can originate from the type of information traders may have at hand. For instance, we could distinguish public versus private information.

discussed in Section 10.3) implies that when the behaviour of the market participants is *free* from any influence of their social (economic / financial) environments in the physics-like manner (i.e. $V = 0$) and when their initial quotes – above and below the previous price – are normally distributed, over numerically equal ranges, the most probable value (i.e. the value of R^2 at $x = 0$) for the future price (at $t = t_{k+1}$) equals the previous price (at $t = t_k$)[3]. Also, it can be figured out that, just like the most probable value, the expectation value of x, that is, $\langle x(t_{k+1})\rangle = \int_{-\infty}^{-\infty} \psi^*(x, t_{k+1})\ \hat{x}\ \psi(x, t_{k+1})\ dx,\ \forall\ t$, is zero as well, implying that $\langle p(t = t_{k+1})\rangle = p(t_k)$, that is the expectation value of the next future price (conditioned on the previous price) is the previous price.

Now, even though both, the most probable value and the expectation value of the future price, as implied by the agent two-preference model for the case of $V = 0$, match with the previous price, in an actual 'price time-series', the next future price practically never comes out to be equal to the previous price. We interpreted this fact as hinting at the need to take into account the environmental influences on the agent-state evolution. We, therefore, considered a generalization ($V = -Kx$) of the *free*-agent two-preference interference model that took into account the influence of the economic / financial environments on the behaviour of the agents (in terms of the evolution of their preferences) and hence, on the distribution of the probability density and on the particle trajectories.

It can be seen from the plots discussed in the Subsections 11.1.2 and 11.1.3 that, when the environmental influence is considered, the shifts in the probability density distributions and also, the bending of the agent trajectories indicate that neither the most probable value nor the expectation value of x would stay at zero, $\forall\ t > 0$. To see this fact in mathematical terms, in the following we give the expression of $\langle x(t)\rangle$ as obtained in [1].

The normalized form of the total wave function at t is given by

$$\psi(x,t) = \frac{\phi_1(x,t) + \phi_2(x,t)}{\sqrt{2\left(1 + e^{-x^2/\rho^2 - m^2(u_1-u_2)^2\rho^2/4\hbar^2}\right)\sqrt{\pi^{1/2}\rho\rho_t}}} \exp\left(-\frac{imK^2t^3}{6\hbar}\right) \tag{11.1}$$

where,

$$\phi_1(x,t) = \exp\left[-\frac{(x - X - u_1t + \frac{Kt^2}{2})^2}{2\rho\rho_t} + \frac{im}{\hbar}(u_1 - Kt)\left(x - \frac{u_1t}{2}\right)\right] \tag{11.2}$$

$$\phi_2(x,t) = \exp\left[-\frac{(x + X - u_2t + \frac{Kt^2}{2})^2}{2\rho\rho_t} + \frac{im}{\hbar}(u_2 - Kt)\left(x - \frac{u_2t}{2}\right)\right] \tag{11.3}$$

[3] It would be interesting to compare and contrast this situation with a free-market economy.

The time-dependent expectation value $\langle x(t)\rangle$ for the particular case of $u_2 = -au_1$ comes out to be

$$\langle x(t)\rangle = \int_{-\infty}^{-\infty} \psi^*(x,t)\,\hat{x}\,\psi(x,t)\,dx = -\frac{Kt^2}{2} + \frac{1}{2}(a-1)u_1 t. \tag{11.4}$$

For $a = 1$, that is, for the case of equal and opposite group velocities, we get $\langle x(t)\rangle = -\frac{Kt^2}{2}$. Thus, for $K > 0$, $\langle x(t)\rangle < 0$ and we can expect the next stock share price to be lower than the previous one. By the same token, a negative value for the K-parameter gives $\langle x(t)\rangle > 0$ and in this case, we can expect the next stock share price to be higher than the previous one.

11.2.3 Interpretations of K and Q

With the already known meaning of x, we propose a crude physics-like interpretation of K by paying attention to the fact that the potential energy-like function is $V = -\mathbf{K}\cdot\mathbf{x} = -Kx$ and therefore, the quantity K is the negative derivative of $V(x)$. Thus, it could be interpreted as a force-like term in the agent two-preference interference situation. To make such a quantity meaningful in the economics/finance context it may be defined as the difference between the supply strength (say, K_s) and the demand strength (say, K_d), that is, $K = K_s - K_d$, by following the 'sign' convention appearing in the plots[4].

This way we can comprehend the shifts in the probability density distributions and the bending of the trajectories for the agent two-preference model situation in the conventional 'demand vs. supply' manner. For instance, when the shift and the bending are in $x < 0$ region, this can be seen as the consequence of the supply force being greater than the demand force (i.e. $K_s > K_d$). This makes $K > 0$ giving $\langle x(t)\rangle < 0$ and therefore, implying that the potential future stock share price will likely be lower than the previous one. This is intuitively appealing as it is well known that 'supply' dominance would tend to decrease the price, whereas 'demand' dominance would tend to rise it. This way, under extreme conditions, when the supply strength is much greater than the demand strength, we can even expect negative future prices!

The interpretation of quantum potential energy seems like both an intriguing as well as an elusive task. Indeed, one can find a variety of interpretations of Q, see, for example, [4], [5], [6]. Therefore, we must admit that a physics-analogous, finance-relevant interpretation of the quantum potential energy-like quantity Q that is being provided here (with reference to the case of 'agents'), is, just like the interpretation provided for K, a crude one. To interpret Q we prefer to proceed in the following manner. Let us recall from Equation (2.42), $[Q = -\frac{\hbar^2}{2m}\frac{\nabla^2 R}{R}]$, that $Q \propto \nabla\cdot\nabla R$, that

[4] By this we mean that for positive values of K the shifts in the probability density distributions are seen in the negative x-direction. The trajectories can also be seen as bending in the negative x-direction.

is, the quantum potential energy arises from the nonzero divergence of the gradient of the amplitude of the agent wave function. Now, the gradient of a scalar field, which itself is a vector field, represents the direction along which the change in the value of the scalar field is maximal. A nonzero (positive) divergence of such a vector field gives its 'outgoingness,' that is, the extent to which its flow (flux) is outgoing with respect to its origin. In light of such an understanding, we can say that 'the *divergence* of the *gradient* of the *amplitude* of the agent's wave function and therefore, Q, represents the outgoingness of the directions along which the wave function amplitude changes maximally.

Quantum potential energy has been considered as the shape-induced internal stress in the wave packet, which is relieved when the wave packet flattens out. This way, it may be seen as a potentiality, providing room for a multitude of evolutions of the initial agent-preferences (i.e. a collection of potentially realizable agent trajectories corresponding to different 'initial condition – final condition' pairs) all subject to the same (i.e., *common-to-all-agents*) social influence V, the classical potential energy-like function.

Assuming an analogy with electrons, it can also be said that the difference between an agent's initial preference (quote) for the future stock share price at which they *would like to trade* and their final choice at which they *would actually end up trading* as displayed by the individual trajectories (in all the trajectory plots) is due to a quantum force-like influence originating from the quantum potential energy-like quantity.

References

[1] von Bloh, K. and Athalye V. 2022. The Double-Slit Experiment in a Uniform Potential. *Wolfram Demonstrations Project*. https://demonstrations.wolfram.com/TheDoubleSlitExperimentInAUniformPotential/.
[2] Donati, O., Missiroli, G. P., and Pozzi, G. S. (1973). An experiment on electron interference. *American Journal of Physics*, 41, 639.
[3] Tavabi, A. H. et al. (2019). The Young–Feynman controlled double-slit electron interference experiment. *Nature*, 9, 10458.
[4] Bohm, D. and Hiley, B. (1993). *The Undivided Universe*. Routledge, Taylor & Francis Group.
[5] Wyatt, R. (2005). *Quantum Dynamics with Trajectories*. Springer.
[6] Fiscaletti, D. (2018). *The Geometry of Quantum Potential: Entropic Information of the Vacuum*. Singapore: World Scientific.

12

Future Directions

In this last chapter of the book, we keep coming back to the potential function and we attempt to connect it to more precise ideas in finance. We also initiate a discussion on agent behaviour and *nonlocality* and causality. Our last words in this book will be centred on what comes next. One key query we have is whether we can consider more complicated real potentials in the two-slit interference experiment with agents (and the agent *two-preference* interference).

In the previous four chapters we discussed how the connections between physics and the economic sciences could be sought and how quantum physics–analogous dynamical models could be systematically constructed in the domain of the economic sciences. Chapter 8 began with an insistence on paying serious attention to the physical (and not just mathematical) foundations of quantum theory when constructing models of the economic-financial reality. By taking a brief review of some of the leading theories from the philosophy of science proposing a variety of criteria for evaluating analogical arguments, it introduced a strategy (described in terms of a four-step procedure) for constructing quantum physics–analogous models. We ended the chapter by explaining the implementation of the first two steps from the procedure. This was done by discussing what a positive analogy could mean at the interface of the domains of physics and the economic sciences. We also stated the hypothetical analogy.

Chapter 9 was about the implementation of the third step from the procedure and hence, for precision, required choosing a specific dynamical situation from the domain of physics and mapping its corresponding elements (objects, relations etc.) with a specific context from the domain of the economic sciences. For this, the electron two-slit interference was chosen and was then mapped with the situation in which individual agents (market participants) would have two alternative preferences: trading at a future stock share price that is higher than and lower than the previous one. In analogy with the electron two-slit interference, this was called the agent two-preference situation. At the end of the chapter we expressed the probability amplitude – that an agent arrives at a potential future stock share price $p(t_{k+1})$, starting from a previous price $p(t_k)$ – in terms of Dirac bra-ket symbols.

For getting a proper quantum physics–like *model* for the 'agents' case, giving a mere symbolic expression for the probability amplitude was not sufficient. What was required were also the dynamical details of the electron two-slit interference case (in terms of the electron wave functions expressed as wave packets), which could be used for the construction of the agent two-preference interference situation in an analogous manner. This was one of the tasks set under the fourth and the last step from the procedure. Also, as another task from the fourth step, clarity and justification of our choice for the interpretation of the wave function and the plausibility arguments for the physics-like interpretations of the agent two-preference interference model parameters were also needed. Chapter 10, therefore, opened up with a discussion on our preference for both, the 'Einstein–Born–Ballentine' ensemble interpretation and the de Broglie–Bohm pilot wave interpretation. By following this combination as our interpretational stance, the chapter presented the dynamical details of the electron two-slit interference case and analogously to the agent two-preference interference case. Chapter 11 provided several plots of the distribution of probability density and the particle trajectories for a variety of potential strengths. As the last piece of our exploration, at the end we gave plausibility arguments for the interpretations of the model parameters for the case of 'agents'.

Of course, this exploration is far from being complete. In fact, it has already given rise to several directions for future work. The purpose of this chapter is to give an outline of some of the future directions that emerge from the previous chapters. Surely, as would befit the typical content of the very last chapter of a book: admonishing (or maybe hedging) words are in order!

12.1 Potential Energy and Agent Heterogeneities

So far, we have been considering the simplest dependence of the potential energy function $V(x)$ on its argument, which is the linear (or equivalently, the 'first power') dependence on x, the particle position. Of course, if we imagine the time-independent function $V(x)$ as a univariate polynomial of degree n, that is, $V(x) = \sum_{i=0}^{n} a_i x^i$, then, in fact, an even simpler case exists with $V(x) = a_0$, which is that of the constant potential energy, given by the first term[1]. However, as we know, the force corresponding to such a potential energy term – to be obtained

[1] Such an expression, of course, excludes the negative powers of x, which could give rise to potentials, for instance, proportional to x^{-n}. We do not deny that these could also serve to model interesting economics-financial influences on the states of the individual agents.

as the gradient of a_0 – will be zero everywhere (i.e. $\forall x$), implying that the corresponding dynamics (motion) will be no different from that for the case of $V(x) = 0$. Therefore, we ignored this case and directly considered the analysis of the dynamics with $V(x) = -\mathbf{K} \cdot \mathbf{x} = -Kx$, that is, the potential energy function having linear dependence on its argument.

However, the constant magnitude K of $\mathbf{K}$ – in other words the position *independence* of $\mathbf{K}$ – means that the force due to the potential energy would influence the motion of all the individual agents, as the elements from the ensemble, in the preference space, in an identical manner, implying an ideal case of the 'behavioural homogeneity' among the agents. Indeed, this can be clearly seen in the plots of the particle trajectories, from the previous chapter, for any given value of K.

Of course, this does not mean that the agent two-preference interference model did not at all consider the heterogeneity in the responses of the agents. Already, by analogously following the electron two-slit interference model, the case of 'agents' did take into account the heterogeneity of the agents at two levels:

1. **The Inter-Slit Heterogeneity:** the first one pertained to the fact that an agent could prefer trading for a future stock share price higher than or lower than the previous price w.r.t. given ranges of quotes. Recalling the analogy between the base preferences in the 'agents' case and the slits in the 'electrons' case, such a heterogeneity may be termed as an *inter-slit* heterogeneity;
2. **The Intra-Slit Heterogeneity:** the second one had to do with the differences in the initial quotes of two agents, even though they could be interested in trading at prices, say, higher than the previous, that is, even though they belong to the same base preference. Such a heterogeneity may be termed as an *intra-slit* heterogeneity.

In terms of the potential energy function $V = -Kx$, just like the case for electrons, we model the influence of the economics-financial environment on the preferences of the individual agents. It could be argued that such an influence is nothing but the agents' interpretation of their economic/financial surroundings based on the information (e.g., the demand–supply scenario at the time of trading) they received. However, with K a constant, V can be understood as a homogeneous influence of the economic/financial environment on all the agents.

As one of the directions for further investigation of the agent two-preference interference model, one could consider a potential energy function, say, of the form $V = -K(x)x$, indicating a position dependence of the force. Again, for simplicity, if the function $K(x)$ is taken as a linear function of x, it will lead to a quadratic potential energy (i.e., $V(x) \propto x^2$), similar to the one influencing the motion of particles in a harmonic manner. Such a preference dependence of the force-like K-parameter will thus represent a heterogenous influence of the social environment on the agents' preferences and hence, will trigger a consequent heterogeneity among the individual agents' responses. With a position-dependent K, that is, $K \equiv K(x)$, it would mean that the information is being differently (subjectively) interpreted by the individual agents and in this way, it may become possible to relate (at least in a simplified manner) the variety in the agents' initial quotes with the variety of (slightly) different cognitive attitudes of the agents.

It would be interesting to develop a proper interpretation of the preference dependence of K. For instance, if the demand and the supply strengths are modelled as *x-independent*, then, as one of the possibilities the function $K(x)$ may involve a third term as a function of x (again, subject to an economic/financial interpretation), thereby becoming a function of x itself.

Alternatively, we could also have the demand and the supply forces themselves be position dependent – implying that each agent arriving at a different initial position interprets the demand and supply scenario in its own way. This way the model may account for the distinction between the public and the private information. This also means that the information interpretation of V would depend upon its form – for the constant K, it is the 'common-to-all' information, influencing all the agents in the same manner, whereas, if K depends on x, it could be argued that the potential function contains the effects not only of public information but also of private information.

12.2 Connections with the Previous Finance Models

As remarked earlier, in forthcoming work, which is somewhat beyond the scope of this book, we shall also consider nonlinear potentials. A reasonable argument can be made that in typical and more advanced social science settings we can (and maybe should) consider nonlinear potentials. We present here two approaches which will be further developed in forthcoming work, within the mechanics of the Bohmian approach.

12.2.1 The Potential Function: the Baaquie Approach

Within the Hamiltonian framework and the double-slit experiment, we would want to argue for the analogy of a potential function in economics. We take a step outside of quantum mechanics in that regard and refer the reader to a highly interesting proposal Belal Baaquie [1] introduced some time ago. In basic economics, the intersection of supply and demand functions determines the stationary prices of goods or services. Baaquie [1] proposes therefore that the so-called microeconomic potential, $V(p)$ can be defined as: $V(p) \equiv De(p) + Su(p)$, where $De(p)$ indicates the demand, whilst $Su(p)$ indicates the supply. In his paper he does show how minimizing this potential leads to the same equilibrium price[2] one finds out of the intersection of supply and demand.

12.2.2 The Potential Approach: Marginal Supply and Demand

In the work by Bouchaud et al. [2], the marginal supply and demand function under the infrequent auctions constraint is introduced. This is in Bouchaud et al., [2], equation 18.6, p. 343. We do not reproduce it here, but for our purposes we are interested in only one part of this function: $\exp\left(-\sqrt{\frac{V}{D}}\left|x - x'\right|\right)$. We note the important parameter D, which we have hitherto defined as per Bouchaud: $D = \frac{1}{2}(\Sigma^2 + \sigma^2 S[.])$. If we define the potential as $V(p) \equiv De(p) + Su(p)$ then we could, for illustrative purposes, set D to some specific value and define the time-independent potential $V(x) = \exp\left(-\sqrt{\frac{1}{D}}x\right) + \exp\left(\sqrt{\frac{1}{D}}x\right)$. In effect, what we are trying to express here is that we can make a case, in a social science context, for both potentials which are linear as well as nonlinear. We do not endeavour to calculate D, as this is beyond the scope of the book. What is interesting here is the following:

1. We can use the idea of Baaquie to define a potential in terms of demand and supply;
2. We can use the parameter D as defined in Bouchaud et al. to construct such a potential; and
3. The tightness of the potential also determines the level of likelihood of the equilibrium price difference occurring at the minimum of the potential.

This approach may lead us to give a more precise proposal for the analogy of the potential function in social science, especially when this is cast within the setting of the double-slit experiment.

12.3 Tenets: From Physics to the Economic Sciences

The new approach for a better comprehension of the economic/financial activities, presented in the previous four chapters, requires the development of models for economics and finance in a physics-analogous way. With such an approach, it becomes inevitable to consider the economic/financial activities from a *causal* viewpoint. Although there is rare use of the terms 'cause' and 'effect' in the formal language of physics, the principle of causality – that for every effect there is

[2] Albeit in a more generalized expression. The microeconomic potential also allows to define how parameters behave in the neighbourhood of the said equilibrium.

a cause – has always been the key tenet driving the developments in physics and also, in other natural sciences and engineering. Therefore, we think that one of the central themes of the future physics-analogous models could be the analysis of the economic-financial processes from a causal viewpoint. This way, it might be possible to obtain interesting, physics-analogous insights into the workings of the world of economics and finance.

In physics, the dynamical analysis of the process in question essentially refers to the cause behind the occurrence of the process, which can be modelled in terms of forces (as usually done in Newtonian mechanics) or in terms of Hamiltonians (as usually done in analytical mechanics and quantum mechanics). Although very rarely stated explicitly, the system Hamiltonian, through the potential energy expression used in the Schrödinger equation, represents the cause underlying its dynamics. With the quantum physics–analogous way of modelling as detailed for the agent two-preference interference case and also, its future extensions – by involving more complicated and perhaps more realistic potential energy functions – as proposed in the last two sections, it seems possible to construct models enabling the analysis of the activities from the world of economics and finance with respect to a causal viewpoint. Of course, the issue of causality in the quantum domain is much more involved than that in the classical domain [3]. Still, it would be intriguing to see how the principle works in a variety of economic/financial contexts.[3]

Apart from the tenet of causality (and hence, from the related but different tenet of determinism), one of the future directions may also involve investigating the economic/financial analogue of a profound feature of quantum physics, namely the existence of nonlocal correlations as observed in quantum experiments in terms of the violation of the Bell-type inequalities and of the Leggett–Garg inequality. The former type of inequalities involve correlations between the measurement outcomes for space-like separated systems, whereas, the latter type of inequalities involve temporal correlations between successive measurement outcomes for a given system. The violation of the Bell-type inequalities is interpreted as the violation of local realism, whereas the violation of the Leggett–Garg inequality is interpreted as the violation of macro-realism. Both these types of inequalities involve *joint probability distributions* for the observables (and thus require experimental determination of joint expectations), as opposed to the specification of *conditional probability distributions* required in the investigation of how an 'earlier' state of one system causally influences the 'later' state of another system.

[3] It may be tempting to give up on causality and simply hold a view that the principle of causality is not applicable in the domain of economics and finance. Still, one will have to examine and explain how this could be the case. In this regard it would also be interesting to investigate how the recently proposed idea of indefinite causal order could be applicable in the economic sciences [4].

As one of the future directions and an extension to the 'causal'-agent two-preference interference model, it would be interesting to investigate the roles of the analogues of the tenets of local realism and macro-realism in the economic/financial contexts.

We constructed the agent two-preference interference model by employing both the Einstein–Born–Ballentine 'ensemble' interpretation and the de Broglie–Bohm pilot wave interpretation of the agent wave function. Nevertheless, quantum physics–analogous models for economics and finance could be constructed by applying the other interpretations of the wave function as well and as one of the future directions, it would be interesting to see what picture of the economic/financial reality the other interpretations provide.

Also, as previously discussed, quantum theory does not just admit a variety of interpretations, but also gives rise to the foundational problems such as the measurement problem and the problem of quantum-to-classical transition. Although the agent two-preference interference model focussed on explaining the distribution of the probability density over future prices, it did not consider how a specific stock share price may pop up out of a distribution over a range of potential prices. With reference to Heisenberg's idea of wave function reduction (or collapse), as another future direction, it would be interesting to shed light on how the reduction of the agent's wave function may take place and a specific stock share price emerges out of it. Also, if the economic/financial reality is described in terms of a quantum physics–like dynamics, then, as yet another future direction, it could be intriguing to investigate, by means of decoherence-like economic/financial models, the conditions under which the agent behaviour may appear classical physics-like. Such investigations may lead to a more balanced and open approach for the construction of physics-like models in economics and finance that would not insist on treating or presuming the behaviour of agents as always quantum physics–like.

12.4 Electrons versus Agents: The Issue of 'Free Will'

The precise analogy exercise leading to a quantum *physics*–like financial model presented in the previous four chapters could open new doors for the investigations pertaining to the nature and the ontology of 'free will'. This complex issue has aspects as diverse as neuroscientific, psychological, philosophical and theological. By virtue of 'free will,' individuals are said to have the ability to make decisions and choices 'freely,' that is, without any external influences.

Recalling the components of analogy discussed in Subsection 8.3.4, it may be argued that there exists a *negative* analogy component between 'the behaviour of *electrons* in the two-slit interference situation' and 'the behaviour of *agents* in the two-preference interference situation', for the reason that the latter type of systems are gifted by nature with freedom of will (and therefore, with freedom of choice), whereas the former type do not possess such an attribute[4]! By freedom of will, we mean the idea of 'spooky free will,' which is often claimed to be incompatible with physics' tenets of causality and determinism.

Now, in physics, the motion of a system is regarded as 'free' when there is an absence of any external potential or force that could influence the motion. In analogy with the 'electron' case, it is interesting to recall that for the 'agent' case (i.e., in the agent two-preference interference context), even with $V = 0$ (i.e. in the absence of any external socio-economic influence, following our interpretation discussed in Subsection 11.2.3), the evolution of the agent preferences was seen to be subject to a nonzero Q (i.e. the quantum potential energy) and the action of a quantum force (as its negative gradient). Thus, even in the so-called (classical) *free* case, the agents' arrival at their final choices of the preferences, when modelled in a quantum physics–analogous manner cannot be regarded as free, especially in the 'spooky' sense as remarked earlier. For an excellent discussion on the neuroscientific distinction between 'spooky' free will and voluntary actions, see [6]. Also, see [7] where Robert Sapolsky scrutinizes the notion of free will with reference to some of the profound scientific concepts such as chaos, emergent complexity and quantum indeterminacy among other things.

This way, the insights into the agent behaviour – such as those offered by the analogous agent two-preference interference model – could be seen to implicitly deny the existence of the spooky free will of the agents participating in the economic/financial activities, thereby rejecting the previously mentioned negative analogy. Also, with reference to our brief discussion in Section 3.8, such quantum physics–like models could pave the way for novel neuroscientific investigations at the interface of physics and the economic sciences. We think that this is one of the most intriguing future directions emerging out of our exploration.

References

[1] Baaquie, B. (2013). Statistical microeconomics. *Physica A*, 392 (19), 4400–4416.

[2] Bouchaud, J. P., Bonart, J., Donner, J. and Gould, M. (2018) *Trades, Quotes and Prices: Financial Markets under the Microscope*. Cambridge, UK: Cambridge University Press.

[3] Riggs P. J. (2009). *Quantum Causality: Conceptual Issues in the Causal Theory of Quantum Mechanics*. Dordrecht: Springer.

[4] Oreshkov, O., Costa, F. & Brukner, C. (2012). Quantum correlations with no causal order. *Nature Communications* 3, 1092.

[4] It is worth mentioning here David Bohm's approach to this issue, which he discusses in [5]. He considers the problem of whether "minds can choose freely among whatever courses of action may be possible" and points to the questionable presupposition about the mind's a priori knowledge on the various alternatives and also, on "which of those is the best."

[5] Bohm, D. (1986). in *Physics and the Ultimate Significance of Time*. Griffin, D. R. (Ed.). State University of New York Press.
[6] Seth, A. (2021). *Being You: A New Science of Consciousness*. New York, NY: Penguin Random House LLC.
[7] Sapolsky R. M. (2023). *Determined: A Science of Life without Free Will*. New York, NY: Penguin Press.

Index

Printed by Integrated Books International,
United States of America